Last-Minute Optics

Optics

**A Concise Review of
Optics, Refraction, and Contact Lenses**

SECOND EDITION

Last-Minute Optics

Optics

A Concise Review of
Optics, Refraction, and Contact Lenses

SECOND EDITION

David G. Hunter, MD, PhD
Ophthalmologist-in-Chief and Richard M. Robb Chair
Children's Hospital Boston
Associate Professor and Vice Chair of Ophthalmology
Harvard Medical School
Boston, MA

Constance E. West, MD
Director, Pediatric Ophthalmology
Associate Professor of Ophthalmology
Abrahamson Pediatric Eye Institute
University of Cincinnati
Children's Hospital Medical Center
Cincinnati, OH

SLACK
INCORPORATED

ISBN: 978-1-55642-927-9

Published by: SLACK Incorporated
                   6900 Grove Road
                   Thorofare, NJ 08086 USA
                   Telephone: 856-848-1000
                   Fax: 856-848-6091
                   www.slackbooks.com

Contact SLACK Incorporated for more information about other books in this field or about the availability of our books from distributors outside the United States.

Library of Congress Cataloging-in-Publication Data

Hunter, David G.
  Last-minute optics : a concise review of optics, refraction, and contact lenses / David G. Hunter, Constance E. West.
-- 2nd ed.
     p. ; cm.
  Includes bibliographical references and index.
  ISBN 978-1-55642-927-9 (alk. paper)
 1. Eye--Accommodation and refraction--Outlines, syllabi, etc. 2. Physiological optics--Outlines, syllabi, etc. 3. Contact lenses--Outlines, syllabi, etc. I. West, Constance E., 1958- II. Title.
  [DNLM: 1. Ocular Physiological Phenomena--Examination Questions. 2. Lenses--Examination Questions. 3. Optical Processes--Examination Questions. 4. Refractive Errors--Examination Questions. WW 18.2 H945L 2010]
  RE925.H76 2010
  617.7'5--dc22
                                                2010009281

Printed in the United States of America.

Last digit is print number: 10   9   8   7   6   5   4   3   2   1

This book is dedicated to David L. Guyton, MD,
who taught us how to have fun with optics,
and who inspired us to share his enthusiasm with our students.

CONTENTS

FIGURE LIST

TABLE LIST

ACKNOWLEDGMENTS

We thank our students, who have reviewed, refined, and critiqued this manuscript over the years.

About the Authors

David G. Hunter, MD, PhD is Ophthalmologist-in-Chief and the Richard M. Robb Chair of Ophthalmology at Children's Hospital Boston. He is also the Vice Chair and Associate Professor of the Harvard Medical School Department of Ophthalmology. Dr. Hunter obtained a bachelor of science in electrical engineering from Rice University and a PhD (in Cell Biology) and MD from Baylor College of Medicine. After he completed an ophthalmology residency at Harvard's Massachusetts Eye and Ear Infirmary, he received fellowship training in pediatric ophthalmology with Drs. David Guyton and Michael Repka at the Wilmer Ophthalmological Institute, Johns Hopkins University. Dr. Hunter lectures on optics and refraction for ophthalmologists-in-training around the world. He is Editor-in-Chief of the *Journal of the American Association of Pediatric Ophthalmology and Strabismus* (2006-2012) and Vice President of the Association for Research in Vision and Ophthalmology (2010). His clinical and research interests focus on strabismus and amblyopia.

Constance E. West, MD graduated from the Massachusetts Institute of Technology with a degree in chemical engineering. Dr. West then received her MD from the University of Massachusetts Medical School. She completed a residency in ophthalmology at Washington University and a fellowship in pediatric ophthalmology at the Wilmer Ophthalmological Institute. She is Associate Professor of Ophthalmology at the University of Cincinnati and Director of the Division of Ophthalmology and the Abrahamson Pediatric Eye Institute at the Cincinnati Children's Hospital Medical Center in Cincinnati, Ohio. Her clinical practice is devoted to pediatric ophthalmology and adult strabismus.

Introduction To First Edition

Optics is dry.

Optics is irrelevant.

Optics is a waste of my time.

I hate optics.

These are the battle cries we have heard from our students over the years as they embarked on yet another annual attempt to review their optics. Fortunately, the cries usually died down by the time we finished our lectures, at least to a more tolerable low-pitched grumbling noise. We believe that this book captures our approach to our lectures on the subject, making optics accessible and understandable, maybe even fun at times, and certainly clinically relevant.

Optics and refraction are more important to the ophthalmologist than ever before. Patients demand rapid visual rehabilitation, and a thorough working knowledge of clinical optics facilitates this. Advances in refractive surgery demand a detailed understanding of refractive error. Insurers and payors now ask for outcome measures—visual acuity and patient satisfaction are often used as the bottom line. Avoiding potential optics and refraction mistakes improves patient satisfaction, which not only gives you the gratification of helping your patients but may also help you hang on to a competitive managed care contract.

This is not a comprehensive treatise on optics. Nor is it a compendium of multiple choice questions. Our goal is to present the most relevant concepts of optics concisely, for those who have a limited amount of time to study. We use a question and answer format to help you identify weak areas, while at the same time reviewing the subject to reinforce areas you already understand. We recommend that you try to answer each question on your own before looking at the answer. To further economize on your study time, if you are most interested in reviewing the most directly clinically relevant material, you should skip past problems that require more extensive calculations. Once you have reviewed what we consider to be the essential concepts covered here, you may want to consult comprehensive texts for a more detailed discussion of areas that interest you, or perhaps an introductory text for areas that still leave you bewildered.

Introduction to Second Edition

We have been pondering the idea of revising *Last-Minute Optics* for years. We would regularly ask ourselves, "What has really changed in clinical optics for the ophthalmologist?" For years the answer has been, "not much." There were no errors to correct. About the only change was that one co-author (DGH) moved to Boston and we both got promoted. Thus, no second edition. But over the years, as we have refined our lectures, we have found that there are a few new areas that should be touched upon—in particular, the continued expansion and refinement of refractive surgery.

In creating this second edition, we have reviewed and revised the entire text while leaving the good stuff alone. There is a new section on refractive surgery, there are some new questions scattered throughout, and we added a new chapter at the end of the book that allows you to synthesize and summarize everything that you have learned. The goal of that last chapter is to help you evaluate patients who present to your office with complaints related directly to optical or refractive concerns.

We hope that you will enjoy this second edition of *Last-Minute Optics*. We have not lost sight of our primary goal—to provide a succinct, targeted study guide in question-and-answer format that will help you review the essentials of optics and refraction. The knowledge gained is not just trivia that might help you pass an exam. It is real-life optics, taken (for the most part) from our experiences with real-life patients (names altered)—an approach that we hope will make you a better doctor.

1

Basic Principles

What is the refractive index of a transparent material?

It is the ratio of the speed of light in a vacuum to the speed of light in that material.

What materials have a refractive index less than 1.000?

Light always goes faster in a vacuum, so no material has a refractive index less than 1.000.

List the refractive index for different parts of the eye and for common lens materials.

See Table 1-1.

Table 1-1	
Refractive Index of Various Materials	
MATERIAL	**REFRACTIVE INDEX (N)**
Air	1.00
Water	1.33
Aqueous	1.34
Vitreous	1.34
Cornea	1.37
Crystalline lens	1.42
PMMA	1.49
Crown glass	1.52
High index lenses	1.6 to 1.8

Hunter DG, West CE.
Last Minute Optics: A Concise Review of Optics, Refraction, and Contact Lenses, Second Edition (pp 1-6).
© 2010 SLACK Incorporated

What is Snell's law?

$$n \sin \varnothing = n' \sin \varnothing'$$

In the formula, n is the index of refraction of a material and ∅ is the angle of a light ray within that material relative to the normal. Plainly stated, Snell's law defines how strongly light is refracted (or bent) when passing from one material to another. Light is bent towards the normal when passing from a medium with a lower index of refraction to a higher one. Think about higher index materials being harder for the light to get through, so the light rays take a shorter path.

Another analogy is to think of a line of soldiers marching on the pavement next to tall grass. As the soldiers at one end begin marching in the grass, they are slowed down, and the line of soldiers is bent toward the normal (Figure 1-1).

Figure 1-1. Marching soldiers mimicking the behavior of light passing from low to high refractive index.

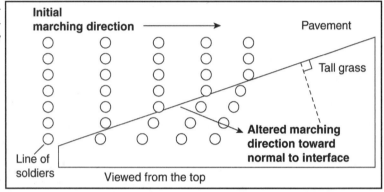

Conversely, light is bent away from the normal when passing from a material with a higher index of refraction to a lower one.

What is critical angle and how does is relate to Snell's law?

When light travels from a material with a higher index of refraction to one with a lower index of refraction, the light ray is bent away from the normal (more parallel to the refractive surface). The "critical angle" is defined as the angle at which light is bent exactly 90° away from the normal. When this occurs, the light ray skims its way along the refractive interface, invisible to an observer on either side. Snell's law is used to calculate the critical angle for any two materials. For example, the critical angle for the interface between glass and air is determined by plugging the numbers into Snell's law: $1.52 \sin \varnothing = 1.00 \sin 90°$. Solving for ∅, the critical angle is 41° for crown glass.

What is total internal reflection and when is it important in ophthalmology?

When the angle of incidence exceeds the critical angle, the light is not refracted, but rather it is *reflected* back into the material with the higher index of refraction—total internal reflection.

In the human eye, light emerging from the angle of the anterior chamber is reflected back into the eye at the tear/air interface, rendering the angle invisible. This phenomenon, total internal reflection, is obviated by gonioscopy lenses. Glass or plastic gonioscopy lenses have a higher index of refraction than tears or the cornea, and the light rays emerging from the angle are indirectly visualized after reflection off of a mirror (4 mirror Goldman gonioprism) or directly visualized with a Koeppe-type goniolens. Total internal reflection is used to our advantage with fiberoptic light sources and in the eyepieces of some indirect ophthalmoscopes.

Note that if the corneal curvature is very steep or the cornea is very large, the light from the angle may strike the cornea at less than the critical angle, and the angle will be visible on direct inspection. This is observed in some patients with keratoconus or marked buphthalmos. A Koeppe lens replaces the cornea/tear/air interface with a plastic/air interface, and at the same time changes the radius of curvature of that interface such that light rays coming from the angle strikes the lens/air interface at less than the critical angle.

Discuss the difference between wave and particle theory of light. Which theory is used in physical optics? Which is used in geometric optics?

No **one** theory can be used to describe the behavior of light, and scientists use different theories, depending on what property of light they are trying to quantify. Physical optics deals with the wave and photon ("particle") properties of light, while geometric optics deals with the behavior of light rays—an artificial, but mathematically useful, construct. When dealing with wave properties of light, the frequency, speed, and wavelength need to be specified. For example, wave theory is used when describing interference and polarization.

Particle, or photon, theory deals with the interaction of light energy with matter. Light is emitted or absorbed in discrete "packets" (photons) of energy. Photon energy is directly proportional to frequency (Energy = [Planck's constant] x [frequency]). For example, fluorescein absorbs higher energy blue light and emits lower energy yellow light.

What fundamental characteristic of light is used to create Haidinger's Brushes and to perform Titmus stereo testing?

Polarization. "Haidinger's brush" is an entoptic phenomenon, seen when a polarizing filter is rotated in front of a blue background. The brush looks like a propeller, centered on the fovea. Titmus stereo testing and other tests of binocular vision (not to mention full color 3D movies) utilize objects that are linearly polarized at 90° to each other. The observer wears glasses with lenses polarized at 90° to each other as well. Each lens allows only the polarized rays from one image to pass; thus, a different image passes through to each eye, and binocular vision is possible.

Vertically polarized light is used to project a letter onto a screen. A polarizing filter is placed in the path of light. What happens to the image when the filter is held horizontally? What about when it is held vertically?

Vertically polarized light will pass through a second vertical polarizing filter virtually unchanged. It is completely blocked by a horizontal polarizing filter. (This assumes the light is linearly polarized, rather than circularly polarized.)

Why do polarizing glasses reduce reflections and glare?

When light is scattered (eg, the blue sky) or reflected (eg, off of the car ahead of you), it becomes partially polarized. The axis of polarization of the scattered light is usually horizontal. Polarized sunglasses use vertical polarizing filters, which block this horizontal component of reflected and scattered light. Some automobile instrument panels, cell phones, and pagers use horizontally polarized displays, which can be difficult to visualize when wearing polarized sunglasses.

What optical phenomena are the basis of antireflection coatings? Name some other ophthalmic exploitations of these phenomena.

Coherence and interference. When two distinct light waves emerge from a single source and then overlap, the peaks of the light waves add to each other to create areas of maximum intensity (constructive interference). In other areas, the peak of one wave matches the valley of the other, and the light is extinguished (destructive interference). This alternation of dark and light areas is known as an interference pattern. Coherence is simply a measure of the ability of two light beams to interfere.

Antireflection coatings are one-quarter wavelength thick. Light rays are reflected both at the air/coating interface and the coating/glass interface. The peak of one reflected wave matches the valley of another, producing destructive interference. That is, if the reflected light wave from one side of the coating hits a "valley" just as the reflected light wave from the other side of the coating hits a "peak," the net reflection will be... nothing. Different coatings are required for different wavelengths. When spectacle lenses have antireflection coatings, the complex coatings required cause the reflections that do occur to have a multi-colored, "rainbow" hue.

Interference filters allow only one wavelength of light to pass by, emphasizing constructive interference at that wavelength. Interference fringe instruments project two beams of coherent light through pinpoint areas of the pupil; the beams overlap on the retina, where they form an interference pattern. Since no image is produced until light reaches the retina, media opacities do not interfere with image formation (as long as the media are 20/200 or better). The ability of the patient to detect a fine interference pattern imaged onto the retina indicates that retinal function is intact.

Why is visual acuity theoretically limited by a small pupil?

Diffraction. Any light ray is bent slightly when it encounters an obstruction or aperture. The smaller the size of the aperture, the greater the spreading of light caused by diffraction. An aperture of less than 2.5 mm (such as the pupil of the eye) begins to cause enough spreading to have an effect on visual acuity.

Why is the sky blue and why is the sun red/orange at sunset and sunrise?

Scattering. Both have to do with the presence of particles—gas molecules—in the atmosphere. Light encounters the gas molecules as it travels through the earth's atmosphere, and *scattering* makes the light "visible" (just like the projector beam in a smoky theater). The intensity of the scattered light is inversely proportional to wavelength to the fourth power, thus shorter (blue) wavelengths are scattered more. When you look up to the sky at noon, lots of blue light is scattered toward your eyes and the sky appears blue. The longer wavelength (red) light continues on a relatively straight path to the earth surrounding you, without being scattered toward your eyes. The sun itself is slightly yellow, having lost mainly its blue components.

At sunrise and sunset, the light reaching you has passed through more of the atmosphere and it has "lost" its shorter wavelength components; you see what is left over—the longer wavelengths (red and orange).

Why don't patients with asteroid hyalosis complain of decreased vision? What would you do if you really needed to visualize the retina of a patient with asteriod hyalosis? (Name three possible approaches.)

Patients are rarely aware of this disorder and it is extremely uncommon that acuity is diminished. The "asteroids" (accumulations of calcium soaps in the vitreous) are small, dense opacities and cast tiny shadows (umbral cones) that reduce the total amount of light reaching the retina. The opacities also scatter a small portion of the light, but they don't reduce acuity because the umbral cones are so short that they don't reach the retina. The ophthalmoscopic appearance of asteroid hyalosis is striking because so much light is reflected back towards the observer by the "asteroids."

Fluorescein angiography is used to examine the retina of patients with severe asteroid hyalosis. Since the yellow-green fluorescent light "source" is inside of the eye, reflections are not a problem for the observer. Scanning laser ophthalmoscopy uses "confocal" optical techniques, which filter out almost all scattered light by precise selection of depth of focus. Ultrasonography, which does not depend on light rays, can also be used.

What are radiometry and photometry? How do they differ? What units are used for each?

Radiometry measures the total power of light (in watts). Photometry measures the power of light in terms of the responsiveness of the eye. Basic units of photometry include candelas, lumens, lux, foot candles, and apostlibs, depending upon how the light is being characterized.

2

Vergence, Lenses, Objects, and Images

What is the definition of vergence? What is a diopter?

Vergence is the amount of spreading of a bundle of light rays coming from a single point (or the convergence of a bundle of light rays heading toward a single point). A diopter is the reciprocal of the distance (in meters) to the point where light rays would intersect if extended in either direction. Vergence is positive if the light rays are converging, and negative if the rays are diverging.

A point source of light is located 25 cm to the left of a + 7.00 D lens. Where will the image of the point source be in focus?

33 cm to the right of the lens.

U (vergence in) + D (lens power) = V (vergence out). U = 1 / 0.25 m = 4 (negative 4 since light is diverging). -4 + 7 = 3. Plus vergence = converging image rays, so the image is to the right of the lens. The reciprocal of the vergence out gives the location of the image (1 / 3 D) = 0.33 m = 33 cm.

Calculate the vergence at positions A through F in Figure 2-1 (answers are found in Figure 2-2).

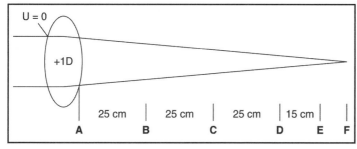

Figure 2-1. Calculate the vergence at positions A through F.

Position A. This is the vergence of light as it leaves the lens. U + D = 0 + 1 = +1.0 (Note that the light rays must intersect 1 / 1 D = 1 m away. This is the key to answering the other parts of the question.)

Position B. This point is 0.25 m from the lens, or 1 - 0.25 = 0.75 m from the point of intersection. Therefore the vergence is 1 / (0.75 D) = +1.33 D. (Plus because the light rays are converging.)

Position C. 1 / (1 – 0.50 m) = +2.0 D

Hunter DG, West CE.
Last Minute Optics: A Concise Review of Optics, Refraction, and Contact Lenses, Second Edition (pp 7-14).
© 2010 SLACK Incorporated

Position D. 1 / (1 − 0.75 m) = +4.0 D

Position E. 1 / (1 − 0.90 m) = +10.0 D

Position F. This is where the light rays intersect. The vergence is infinite.

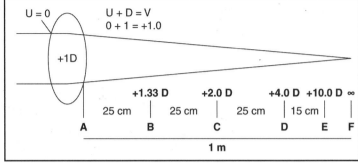

Figure 2-2. Solution to Figure 2-1.

Calculate the vergence in the image space at positions A through F in Figure 2-3 (answers are found in Figure 2-4).

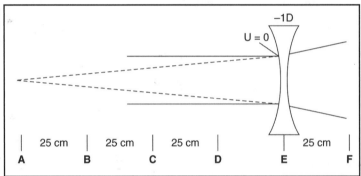

Figure 2-3. Calculate the vergence in the image space at positions A through F.

Use the vergence formula to calculate the answers: U + D = V. The vergence leaving the lens is (0 + -1) = -1. Therefore, the rays of light intersect (1 / 1 D) = 1 m to the left of the lens. Point A is at the point of intersection.

Position A. Infinite vergence

Position B. (1 / 0.25 m) = -4.0 D (minus because it is divergent light)

Position C. (1 / 0.50 m) = -2.0 D

Position D. (1 / 0.75 m) = -1.33 D

Position E. -1.0 D. This is determined by U + D = V (0 + -1 = -1).

Position F. (1 / 1.25 m) = -0.8 D

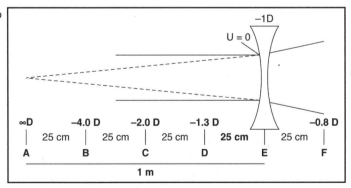

Figure 2-4. Solution to Figure 2-3.

Calculate the vergence at positions A, B, and D in Figure 2-5. What is the value for C? (Answers are found in Figure 2-6.)

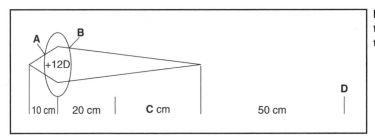

Figure 2-5. Calculate the vergence at positions A, B, and D.

Position A. -10 D. The vergence entering the lens is (1 / 0.1) = -10 (minus for divergent light).

Position B. +2 D. U + D = V. -10 + 12 = +2.

Position C. 30 cm. The light rays emerging from the lens intersect at (1 / 2) = 0.5 m = 50 cm from the lens. Therefore, "C" + 20 = 50 cm, so C is 50 – 20 = 30 cm.

Position D. -2 D. The light rays are diverging (minus vergence) and are located 0.5 m from the point of intersection. (1 / 0.5 = 2.0).

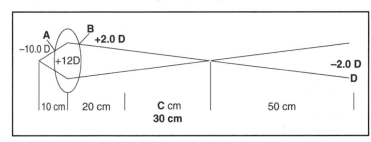

Figure 2-6. Solution to Figure 2-5.

In Figure 2-7, locate the image for the following values of X (answers are found in Figure 2-8).

A. X = 100 cm
B. X = 50 cm
C. X = 250 mm
D. X = 12.5 cm
E. X = 0.011 m

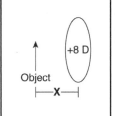

Figure 2-7. Locate the image for the values of X given in the text.

A. 14 cm. U = 1 / 100 cm = 1 / 1 m = -1 D (divergent light). U + D = -1 + 8 = +7.
   1 / V = 1 / 7 = +0.14 m = 14 cm to the right of the lens.

B. 16.7 cm

C. 25 cm

D. Infinity. -1 / 0.125 m = -8. U + D = -8 + 8 = 0. Zero vergence = parallel rays of light.

E. -1.2 cm. 1 / -0.011 = -91. U + D = -91 + 8 = -83. 1 / -83 = -0.012 m = -1.2 cm (1.2 cm to the left of the lens.)

Recalculate the answers in the problem above if the lens in Figure 2-7 had a power of -8 D instead of +8 D.

 A. -11.1 cm (-1 + -8 = -9; 1 / 9 = 0.111 m = 11.1 cm to the left of the lens)
 B. -10.0 cm
 C. -8.3 cm
 D. -6.25 cm
 E. -1.0 cm

Figure 2-8. Locate the image for the given lens combinations.

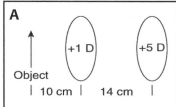

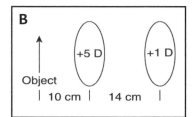

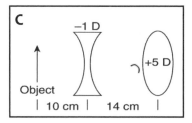

How far is the image from the object in the lens systems shown in Figure 2-8A, B, and C?

These problems can be answered by treating the lenses one at a time, sequentially, from left to right. The image of the first lens ("Image 1") is the same as the object of the second lens ("Object 2"). The first lens is then ignored when locating the image produced by the second lens ("Image 2").

 A. 1.24 m to the right of the object. In Step 1 of Figure 2-9, U + D = V gives a vergence of -9 D leaving the first lens. This means that Image 1 is located (1 / 9) = 0.11 m = 11 cm to the left of the first lens. Image 1 (Step 1) becomes Object 2 (Step 2). Therefore, for Step 2 (Figure 2-9), Image 1 (or Object 2) is located (11 + 14) = 25 cm from the second lens. The vergence entering the second lens is (1 / 0.25 m) = -4 D. U + D = V gives a vergence of +1 leaving the second lens. The image of the second lens is located at (1 / +1) = 1 m to the right of the second lens, or 1.24 m from Object 1.

Figure 2-9. Solution to Figure 2-8A.

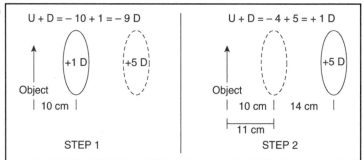

B. 29 cm to the left of the object (Figure 2-10). The image of the first lens is located 20 cm to its left. This image ("Image 1") becomes the object of the second lens ("Object 2," Step 2). The distance from Image 1 (which is Object 2) to the second lens is (20 + 14) = 34 cm. "U" for the second lens is (1 / 0.34) = -2.9 D. The vergence leaving the second lens is (-2.9 + 1) = -1.9 D. Image 2 is (1 / -1.9) = -0.53 m or 53 cm to the left of the second lens, or (53 - 24) = 29 cm to the left of the original object.

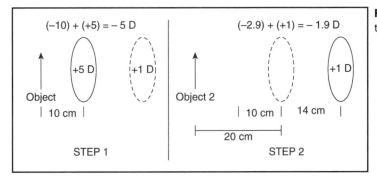

Figure 2-10. Solution to Figure 2-8B.

C. 1.64 m to the right of the original object (Figure 2-11). Image 1 (or Object 2) is 9 cm to the left of the first lens or (9 + 14) = 23 cm to the left of the second lens. The vergence leaving the second lens is therefore (-4.3 + 5) = +0.7 D. The image is 1 / +0.7 = 1.4 m to the right of the second lens, or (1.4 + 0.14 + 0.10) = 1.64 m to the right of the original object.

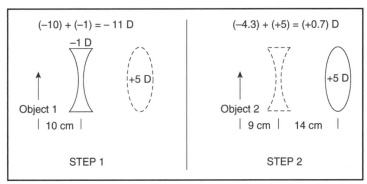

Figure 2-11. Solution to Figure 2-8C.

You are eating a peanut butter and jelly sandwich and drop a large, perfectly smooth, hemispherical glob of apple jelly on your glasses. The jelly has an index of refraction of 1.33. The radius of curvature of the glob is 25 mm. What is the refractive power of the surface where jelly meets air?

Use the formula,

$$D_S = \frac{(n' - n)}{r}$$

where the refracting power is in diopters of a spherical surface. In this case, n' = 1.33, n = 1.00 (refractive index for air), and r = 0.025 m. Thus, the power is 13.2 D. It is positive power because the medium with the higher refractive index has a convex surface.

Before leaving for a scuba diving trip in Hawaii, you pack a +20 D glass lens in your suitcase. While scuba diving, you decide to use your lens to get a closer look at the scales of a most interesting fish. What is the power of the +20 D lens under water? Assume you are holding an ideal thin lens.

From an earlier section, you know that the index of refraction of air is 1.00, glass is 1.52, and water is 1.33. The refracting power of a thin lens is proportional to the difference in refractive index between the lens and the medium. Since the radius of curvature does not change, it does not need to be included in the equation.

$$\frac{D_{air}}{D_{water}} = \frac{+20}{D_{water}} = \frac{n_{lens} - n_{air}}{n_{lens} - n_{water}} = \frac{1.52 - 1.00}{1.52 - 1.33} = 2.74$$

Therefore, $D_{water} = D_{air} / 2.74 = +7.3$ D. (The lens would work only as a 1.8X magnifier in water (assuming a 25 cm reference distance underwater), rather than as a 5X magnifier in air—see magnification section following to see how to calculate this.)

How does one distinguish between real and virtual images when evaluating lens systems?

A real image is always on the same side of the lens as the actual rays of light forming the image (the image rays). A virtual image must be located by *imaginary extensions* of the light rays into object space. The same is true of real and virtual objects.

What is the focal length of a +5 D lens? Where is its secondary focal point? What does that mean, "secondary focal point?"

Focal length is the distance from the ideal thin lens to each of its focal points. Focal length in meters is the reciprocal of the power of the lens in diopters. In this case, the focal length is $1 / 5$ D = 0.2 m or 200 mm. When an object is placed at the primary focal point of a lens, rays emerging from the lens will be parallel. Parallel incident rays are brought to focus at the secondary focal point of a lens. The secondary focal point of the lens above is 200 mm (20 cm) to the right of the lens, assuming the light is coming from the left.

What is the geometric center of a lens? What about the center of gravity? The optical center? How do you find each? How are these points related to each other?

The geometric center of a lens is the center of an imaginary rectangle enclosing the lens (Figure 2-12). The center of gravity of a lens is the point at which the lens will balance on your pencil (this is one way to locate it, if you have superhuman patience and steadiness). The optical center of a lens is the point at which the lens has no net prismatic effect on the light rays it refracts. To locate the optical center of a lens that has no ground-in prism, place the lens in a lensmeter and adjust the focus until you can identify the center of the lensmeter target. As you move the lens horizontally or vertically, you will see the lensmeter target move. Adjust the position of the lens until the center of the target is in the center of the eyepiece cross hairs. You have located the optical center of the lens. The optical, geometric, and gravitational centers of lenses have absolutely nothing to do with each other. Only the optical center is of importance to ophthalmologists.

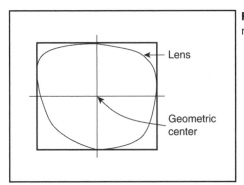

Figure 2-12. The geometric center of a lens is not important to ophthalmologists.

3

The Model Eye

What is shown in Figure 3-1 (answers are found in Figure 3-2)? Fill in the missing numbers. Name the most well-known version of this.

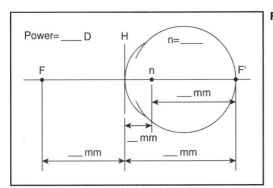

Figure 3-1. Fill in the missing numbers.

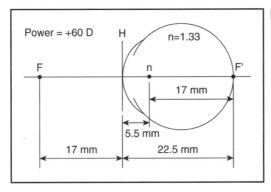

Figure 3-2. Solution to Figure 3-1.

The *reduced schematic eye* is used in many optics problems because it is easier to use than more complicated models, while still being accurate enough for almost all of the calculations we perform. It has only one surface, one nodal point, and one equivalent (principal) plane. The dioptric power is 60 D, the axial length is 22.5 mm, the anterior focal length is 17 mm measured from the cornea, and the nodal point is 5.5 mm posterior to the cornea. Note that this places the nodal point 17 mm in front of the retina, and this number is useful for calculating image size on the retina (see the Goldman perimetry problem). The index of refraction is 1.33.

Hunter DG, West CE.
Last Minute Optics: A Concise Review of Optics, Refraction, and Contact Lenses, Second Edition (pp 15-18).
© 2010 SLACK Incorporated

There are a number of other models of the eye's optical system, the most famous of which was developed by Gullstrand, the only ophthalmologist (so far) to win a Nobel Prize. His schematic eye is somewhat more complicated than the reduced schematic eye, as it has two principal planes and nodal points. We don't believe it is important to memorize the numbers in the Gullstrand model; just know how it differs, in principle, from the reduced schematic eye.

In the reduced schematic eye, why doesn't the nodal point coincide with the principal plane?

The higher refractive index inside of the eye "pulls" the nodal point posteriorly.

If the power of the reduced schematic model eye is +60 D, and if refraction "occurs" at the principal plane (H), why isn't the retina located 1 / 60 = 17 mm behind the principal plane? Why should I care about the principal plane?

While the anterior focal point of the eye is measured from the principal plane, the *posterior* focal point, which coincides with the retina, is measured from the nodal point. Thus, the *anterior* focal point is (1 / 60) = 17 mm in front of the principal plane, while the posterior focal point is 17 mm behind the *nodal point*, which in turn is 5.5 mm behind the principal plane. You should care about the principal plane if you ever do any ray tracing, or if you need to know the vergence of light entering the eye from a near object.

During Goldmann perimetry, a test spot is viewed by an emmetropic patient. How many times larger is the height of the test spot on the perimeter, compared to its height on the emmetrope's retina?

The tangent bowl is 33 cm (330 mm) from the patient's cornea. Using the reduced schematic eye and similar triangles,

$$\frac{\text{Object}}{\text{Image}} = \frac{330}{17} = 19.4$$

or about 20 times larger (Figure 3-3).

Figure 3-3. Object and image height in Goldmann perimetry.

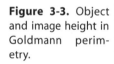
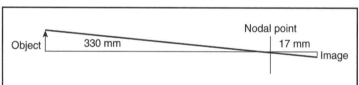

Note that technically, we should add the 5.5 mm distance from the cornea to the nodal point of the eye; however, as long as the object is distant from the eye, the 5.5 mm is not significant compared with the distance from the eye to the object. (It would give us 335.5 / 17 = 19.7).

Assuming a disk diameter of 1.7 mm, what is the diameter of the blind spot when plotted on a tangent screen 2 meters from the eye?

Consider the diagram in Figure 3-4.

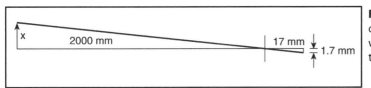

Figure 3-4. Diameter of the blind spot when plotted on a tangent screen.

Using similar triangles, 1.7 / 17 = x / 2000. Rearranging, x = (1.7 / 17) 2000 = 200 mm, or 20 cm. In general, we can use the formula:

$$\frac{\text{Object height}}{\text{Retinal image height}} = \frac{\text{Distance from nodal point}}{17 \text{ mm}}$$

When silicone oil is placed in a normal phakic eye, what happens to the refractive error while the silicone oil remains in the eye?

The first three surfaces maintain their refractive contributions to the total refractive power of the eye, but the contribution of the rear surface of the crystalline lens changes. First consider the normal phakic eye: the refractive index of the crystalline lens is 1.42, and that of the vitreous is 1.33. Because the surface with the higher refractive index is convex, the posterior surface of the lens has plus power.

After vitrectomy with silicone oil, the vitreous is replaced by silicone oil with a refractive index of 1.4034. Thus, the posterior surface still has plus power, but substantially less than prior to silicone oil – there is less of a change in refractive index at the refractive surface. A typical eye will lose 5 to 7 D of power as a result, giving the eye a hyperopic shift.

When silicone oil is inserted into an otherwise-normal aphakic eye, what happens to the refraction while the silicone oil remains in the eye?

A typical aphakic eye that would normally have 10 to 12 D of hyperopia since the refractive contribution of the crystalline lens has been lost.

Normally, the rear surface of the cornea has minus power. However, when silicone oil is in contact with the posterior surface of the cornea, the rear surface of the cornea has plus power due to the greater refractive index of silicone oil compared to the cornea (1.4034 versus 1.37). This gain in plus power partially compensates for the loss of the crystalline lens, and the resultant refractive error is about 4 to 6 D of hyperopia.

Discuss the refractive considerations when silicone oil and intraocular lenses must be used together.

Consider first that if an eye has a cataract and lens extraction with IOL is indicated, then the IOL material, design and power should be chosen carefully if silicone oil is to be used. Lens style and material should be considered first. Silicone lenses should be avoided at all costs in eyes that have, or may require, silicone oil. A convex-plano or meniscus style lens (where all, or most, of the power of the IOL is on the anterior surface of the lens) should be selected in order to minimize changes in lens power after removal of silicone oil from the eye.

Another important consideration is making sure ultrasound measurements are accurate. If ultrasound biometry is used, corrections must be made to compensate for the slower speed of the sound waves in silicone oil. Some instruments will make this correction with the press of a button, while with other instruments the correction must be made manually.

Once accurate axial length measurements are obtained, the IOL power calculations should be modified to take into account the increased index of refraction of silicone oil (1.4034) compared to vitreous (1.33). In normal sized eyes, about 3 D should be added to the calculated power of the IOL.

Visual Acuity Testing

What is the optimal size for a pinhole used to measure "pinhole" acuity?

The optimal size is 1.2 mm, as larger pinholes do not effectively neutralize refractive error and smaller pinholes markedly increase diffraction and decrease the amount of light entering the eye. When testing eyes with larger refractive errors (>5D), it is necessary to correct the majority of the refractive error with a lens in order to obtain best potential vision using a pinhole.

What are the three general types of visual acuity? What is the normal threshold of each, in seconds of arc?

Minimum visible: The detection of the presence or absence of a black dot on a white background—1 to 10 seconds of arc.

Minimum separable: Often called minimum resolvable or minimum legible, or ordinary visual acuity. The detection of the identifying features of a visible target—30 to 60 seconds of arc.

Spatial minimum discriminable: Hyperacuity (eg, vernier acuity). Determination of the relative location of 2 or more visible features relative to one another (eg, a break in a line)—3 to 5 seconds of arc.

What are the dimensions of a 20 size optotype "E" from a chart that is meant to be viewed at 20 feet? How much visual angle does it subtend? What about the same size Landolt "C"?

A 20/20 "E" from a chart meant to be viewed at 20 feet is about 9 mm tall, and each leg and space is about 2 mm tall. The "E" subtends 5 minutes of visual angle; each leg and space subtends 1 minute of visual angle.

A Landolt "C" also subtends 5 minutes of visual angle, is about 9 mm tall, and the opening in the "C" is about 2 mm (1 minute of arc).

How is near vision quantified?

Reduced Snellen acuity (inches or cm), Jaeger or Point type *at a noted distance,* usually 14 inches, with notation about any correction worn.

Hunter DG, West CE.
Last Minute Optics: A Concise Review of Optics, Refraction, and Contact Lenses, Second Edition (pp 19-22).
© 2010 SLACK Incorporated

What are some of the methods used to test visual acuity in preliterate children?

- Blinks to light
- Responds to optokinetic nystagmus stimulus
- FFM/CSM (fixes, follows, maintains or central, steady, maintained)
- Grating acuity cards (Teller acuity cards, preferential looking)
- Picture optotypes (Allen figures, LEA symbols, Kindergarten chart)
- Letter optotypes (HOTV, illiterate E, Landolt C)
- VEP (visual evoked potential)

  Note that only VEP requires no motor response whatsoever from the subject being tested.

What are some of the methods to test visual acuity in illiterate adults?

- Numbers
- Tumbling or illiterate E
- Landolt C's
- Grating acuity

Name some factors other than disease that reduce measured visual acuity.

- Uncorrected ametropia
- Eccentric viewing
- Decreased contrast
- Large (>6 mm) or small (<2.5 mm) pupil size
- Young or old age
- Large pupils reduce visual acuity due to increased coma and spherical aberration. Smaller pupils reduce these optical aberrations, but at the expense of increased diffraction.

What is the definition of legal blindness in the United States?

Although the visual acuity requirements for a driver's license vary from state to state, the definition of legal blindness does not. A patient is legally blind when the best corrected vision in the better eye is 20/200 or worse, or when, despite the acuity obtained, the field of vision in the better eye is 20° or less in diameter.

Tell me about the chart used in the Early Treatment of Diabetic Retinopathy Study (ETDRS). Why is it used in large studies that use visual acuity as an outcome?

The ETDRS (or Ferris-Bailey) distance visual acuity charts use sans serifed Sloan optotypes of equal viewing difficulty (not all letters in the alphabet are of equal difficulty). Each line has five letters, and the space between the letters is equal to the letter size on that line. The progression of optotype heights is geometric, decreasing in 0.1 log unit increments. A three line (15 letter) decrease or increase in acuity anywhere on the chart corresponds to a doubling or halving, respectively, of the visual angle, making mathematical comparisons of visual acuities simpler and psychophysically correct (Figure 4-1).

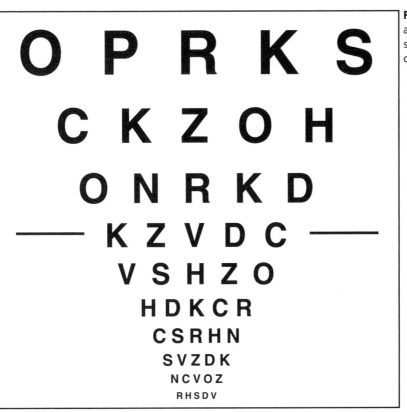

Figure 4-1. Appearance of a Ferris-Bailey style of visual acuity chart.

How is visual acuity quantified when the patient is unable to see a standard Snellen chart?

The most accurate way is to move the patient and chart closer together, noting the size of the optotype seen at a certain distance. Because the usual office charts jump from 20/100 to 20/200 to 20/400, moving the patient and the chart closer together allows a more accurate assessment of the patient's vision. A patient who reads 30 size optotype at 5 feet (5/30) has 20/120 vision at 20 feet, but you would underestimate his or her vision, recording a visual acuity of 20/200 with standard testing charts. Counting fingers at a certain number of feet, hand motions, or light perception with or without projection may suffice when more accurate methods are not available.

How is contrast sensitivity evaluated?

Contrast sensitivity is assessed with the best correction in place and with normal room illumination. Pupils should be in their natural (undilated) state. The patient is presented with sine wave gratings with progressively decreased contrast (Vistech chart). The lowest point (threshold) at which the grating orientation can be identified accurately is determined. The results are presented on a chart that shows the threshold for a number of different grating spacings. The test can also be performed using specially designed letter optotypes (Pelli-Robson chart).

5

Refraction and Optical Dispensing

What is the difference between axial and refractive myopia?

In axial myopia, the refractive power of the eye is normal (about 60 D), but the eye is too long. In refractive myopia, the refractive power of the eye is too strong (more than 60 D), while the length is normal. Both situations create a focal point in front of the retina (Figure 5-1).

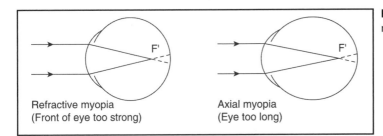

Figure 5-1. Types of myopia.

Is there a difference between refractive and axial hyperopia?

Yes, in axial hyperopia, the refractive power of the eye is normal (about 60 D), but the eye is too short. In refractive hyperopia, the refractive power of the eye is too weak (less than 60 D), while the length is normal (aphakia is the extreme example). Both situations move the focal point behind the retina (Figure 5-2).

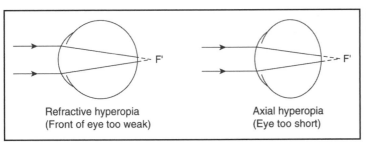

Figure 5-2. Types of hyperopia.

Hunter DG, West CE.
Last Minute Optics: A Concise Review of Optics, Refraction, and Contact Lenses, Second Edition (pp 23-30).
© 2010 SLACK Incorporated

A patient presents with a change in refractive error of -2.00 D over the past 4 months. What are some possible causes of acquired myopia? Consider all age groups.

Acquired refractive error is a subject where a detailed knowledge of the differential diagnosis can help you detect conditions that might be vision- or even life-threatening. Acquired myopia can be divided into refractive and axial causes.

Things that make the refractive power of the eye greater:

- Lens nucleus or shape changes (increased curvature or increased refractive index)
    - o Cataract
    - o Diabetes
    - o ROP
    - o Lenticonus
- Lens repositioning (increased effective lens power)
    - o Ciliary muscle shift
        - ▲ Toxemia of pregnancy
        - ▲ Drugs (topiramate, chlorthalidone, sulfonamides, tetracycline, carbonic anhydrase inhibitors)
    - o Lens movement
        - ▲ Anterior lens dislocation
        - ▲ Peripheral dislocation, bringing the relatively higher power of the lens periphery into the path of light
- Ciliary muscle tone (increased lens curvature)
    - o Antihistamines
    - o Miotics
    - o Excessive accommodation (law and medical students)
    - o Improper refraction technique (inadequate fogging)
- Corneal power increase (increased curvature)
    - o Keratoconus

And things that make the eye longer:

- Congenital or developmental glaucoma
- Posterior staphyloma
- Idiopathic progressive myopia
- Scleral buckle

Please discuss some causes of acquired hyperopia.

This can also be divided into refractive and axial causes.

Refractive causes (things that weaken the refractive power of the eye):

- Absent lens
    - o Aphakia
    - o Severe dislocation of the lens
- Repositioned lens
    - o Posterior dislocation of the lens
- Weak accommodation
    - o Tonic pupil (Adie's syndrome)
    - o Third nerve palsy

      o Trauma

      o Drugs (chloroquine, phenothiazines, antihistamines, benzodiazepines, marijuana)

Axial causes (things that push the retina forward):

- Central serous retinopathy
- Choroidal tumor (choroidal melanoma or retinal hemangioma)
- Orbital mass with pressure on the posterior globe

What are some causes of congenital or acquired astigmatism?

Astigmatism may be secondary to:

External factors

- Masses (eg, dermoid, lid tumors, chalazia, anterior orbital tumors)
- Ptosis

Corneal or corneoscleral causes

- Keratoconus
- Pterygium
- Pellucid marginal degeneration
- Terrien's
- Simple corneal astigmatism

Lenticular causes

- Lenticonus
- Lens decentration (dislocation—eg, Marfan's)
- Ciliary body tumor
- Lens coloboma
- Simple lenticular astigmatism.

Note: retinal disease does not cause astigmatism, only distortion and non-astigmatic blur.

Monte McGoo is a 22-year-old draftsman with a refraction of -22.00 sphere in both eyes. He is concerned about the weight and appearance of his very thick glasses. What interventions might you offer?

First, make sure the patient really needs all that minus; perform a cycloplegic refraction. Of course, you could discuss the options of contact lenses or refractive surgery, but if the patient cannot pursue these options there are ways to make the glasses lighter or thinner. Most important is to recommend that the lenses be made of high-index material—the higher refractive index means that the same power can be achieved with less lens curvature and thus less thickness at the edge of the lens. Smaller frames can help, since the lenses get thicker as the distance from the optical center increases.

A few other interventions are more nuanced but may be helpful in some cases: If the front of the lens has a steep base curve (which essentially adds plus power), it requires that more minus be removed from the back of the lens; a flatter base curve can reduce this problem. The optician can bevel and polish the edges of the lenses, and can help the patient choose frames that hide the thick edges.

A 27-year-old woman, whom you have painstakingly refracted, returns complaining of trouble with her distance vision, especially at night. You repeat your refraction and obtain the exactly the same results. What are some probable causes of the patient's complaints?

A combination of factors may contribute to night myopia:

- When the pupil dilates at night because of lower light levels, rays striking the crystalline lens' periphery are refracted more strongly (positive spherical aberration). This causes the image to be moved forward into the vitreous, effectively producing myopia. The larger pupil may also reveal irregular astigmatism.

- Refraction lanes are less than infinitely long, and patients are therefore refracted for 6 meters (20 feet). This leaves the patient, whether hyperopic or myopic, 1/6 D (0.167 D) undercorrected. Add a little minus (-0.25 D) to avoid this iatrogenic cause.

- In lower light levels, spectral sensitivity shifts to shorter wavelengths (Purkinje shift), and chromatic aberration (see later discussion) moves the focal point anteriorly, also producing myopia.

- When it is dark and there are no accommodative targets to "anchor" accommodation, the patient may "fog" herself simply through overactive accommodation.

Describe some problems with high plus-powered spectacles.

- A ring scotoma is seen with high plus spectacles (as in aphakia). As the patient looks laterally to see an object, the object seems to disappear. The roving ring scotoma moves centrally as the eyes gaze to the side. Aphakic patients thus feel that they have no reliable peripheral vision. This artifact is a consequence of the prismatic effect of lenses.

- Pincushion distortion (periphery of image is magnified more than center)

- Excessive magnification

- Weight

- Cost

A patient complains that his glasses fog and get dirty. Discuss some possible causes and remedies.

The frames may be too tight, with poor air circulation; adjust tilt to minimize contact and increase vertex distance. Some frames are so large that they rub the eyebrows or rest on the cheeks—change to a smaller frame size. If the glasses fog frequently (eg, butchers might complain of this when they go in and out of the meat locker), change to plastic lenses which conduct heat more poorly than glass. Medical conditions may also contribute to the problem: epiphora may cause fogging and seborrheic dermatitis may promote deposition of dirt and oil on the lenses.

After a careful refraction and in-office trial, your father (an electrical engineer) spends $475 on his first new glasses in 15 years. As soon as he receives the new glasses, he calls, complaining, "My eyes feel like they are being sucked out of my head." What should you do?

This is one of many colorful ways that patients use to describe their response to subtle changes in lens design or positioning. Check location of optical centers for induced prism, check segment height, frame alignment, compare the base curve of the new lenses to that of his old lenses, and look for front versus back cylinder lenses. See also Section 18, *Good People, Bad Optics: The Dissatisfied Patient.*

After refracting your best referral source as a favor, she returns with her new glasses, complaining that her vision is not as sharp as it was in your office at the end of the refraction. What are some of the causes and remedies?

Check that the prescription was filled correctly. Is there an uncompensated change in vertex distance? Perhaps the pantoscopic tilt of trial and dispensed glasses are different, or you may have refracted her incorrectly, or failed to correct vision to "infinity." See also Section 18, *Good People, Bad Optics: The Dissatisfied Patient.*

A patient calls complaining that the duplicate pair of glasses you prescribed is not as comfortable as the original pair; what factors do you consider?

Error in lens power; change in optical center separation, cylinder form (front versus back toric), frame size, pantoscopic tilt, base curve, and segment type/height. See also Section 18, *Good People, Bad Optics: The Dissatisfied Patient.*

Mr. Mom calls, stating his 18-month-old son just won't keep the glasses you prescribed on his face. What might you suggest?

Make sure Mr. Mom understands how important the glasses are to the development and maintenance of his son's visual acuity and alignment. Mr. Mom may not want his child to wear glasses, and either he is not supporting the child's efforts to wear glasses or he may be looking for an excuse to discontinue the glasses. Opticians often fit small children and infants with frames that are too big for their faces, especially a child with a flat nasal bridge. Be sure to suggest an optician who has an interest in, and expertise fitting, children's glasses. Children are tough on glasses, and parents should expect to have the frames adjusted frequently.

If a hyperopic child will not wear his glasses despite encouragement, consider a short course of cycloplegia to relax his accommodative efforts and make him more appreciative of the improvement in acuity. If he still doesn't appreciate the difference and glasses wear is important, you might consider a short trial of arm splints, or if possible wait a few months and try again. Phospholine iodide is another option, especially if you are using the glasses to treat a refractive or accommodative strabismus.

Mrs. Whiner presents with two pairs of bifocal glasses, complaining, "I can't read with the new pair you gave me, doctor." Her old prescription is +2.00 -2.00 x 090 OU, with +2.50 add. Her new pair is +0.75 -1.50 x 090 OU, with +2.50 add. What happened?

You forgot to adjust the add when you added more minus spherical equivalent in the distance prescription. The spherical equivalent of the old near prescription is +3.50 D, and the spherical equivalent of the new pair is +2.50 D. Her new pair gives her 1 D less add for near. In other cases, patients will have trouble when you *increase* the add (giving them much better visual acuity) because it shortens the working distance, but in this example you have increased the working distance. Also remember to check the segment type and height in the new glasses. See also Section 18, *Good People, Bad Optics: The Dissatisfied Patient.*

A 24-year-old professional baseball player is contact lens intolerant. He complains of severe glare and reflections with his new glasses. The old glasses were just fine. Why has he developed this new problem?

If there is no ophthalmic cause for his photophobia, he may be bothered by too much illumination or multiple sources of illumination, such as the floodlights used in baseball fields. He may be telling you he is noticing reflections from the lens surfaces. You could try to change the angle between the light source and the eyes by suggesting that the optician shorten the vertex distance or change pantoscopic tilt. One of the easiest solutions is to consider an anti-reflective coating. See also Section 18, *Good People, Bad Optics: The Dissatisfied Patient.*

Why might a patient dislike high-index lenses?

Although high-index lenses are thinner and of a lighter weight than conventional plastic lenses, they have more chromatic aberration, especially in higher powers. When a white spot is viewed away from the optical center, it will have a blue or yellow "ghost" or shadow. Also, the area of the lens giving the clearest vision is smaller.

A 49-year-old woman just received her first pair of bifocals, and complains of inability to walk without stumbling when wearing bifocals. What might you consider when evaluating this patient?

Some patients just don't want to wear bifocals and will use any and all excuses to get rid of them. Some will even go without correction, despite very poor visual acuity. Executive bifocals may be particularly difficult to adjust to, because the add occupies the entire lower visual field. Advise the patient to flex her neck to look where she is stepping, rather than looking downward through her bifocals. She may do best with separate reading and distance correction, or with single vision "driving and walking" glasses and separate bifocal "reading" glasses.

Although progressive addition lenses offer a cosmetic improvement, they induce a good deal of irregular astigmatism outside the central channel of clear vision. In addition, objects will seem to "swim" as the patient turns her head from side to side.

During residency you learned to retinoscope at a working distance of 67 cm from the patient, but you had an accident recently and can't extend your arms as far. Now you perform retinoscopy at a distance of 50 cm. How do you modify the value obtained at neutralization to obtain the refraction at infinity in each case? Why?

Before your accident you subtracted 1.50 D (that is, you modified the value obtained at neutralization by 1.50 D in the minus direction), and afterwards you should subtract 2.00 D. At neutralization, the patient's eye is in focus at the peephole of the retinoscope (ie, the far point of the patient's eye is at the peephole of your retinoscope), which was 0.67 m (before the accident) or 0.5 m (after the accident) away. To move the far point of the eye to infinity, 1 / 0.67 m = 1.50 D was subtracted from the refraction obtained at neutralization before your accident, and 1 / 0.5 m = 2.00 D should be subtracted after your accident.

A 25-year-old accountant presents to your clinic for refraction after losing his glasses. He had early bilateral congenital cataract extractions in infancy; followed by corneal transplants for keratoconus; complicated by HSV keratitis, long-standing allergic conjunctivitis, and constipation. One pupil is large and irregular, the other is 4 mm and irregular. Uncorrected visual acuity is 20/400 OU, but it improves to a brisk 20/30 in each eye through a pinhole. Retinoscopy is not helpful due to difficulty viewing the retinoscopic reflex. What do you do?

- Concentrate as much as possible on the central portion of the light reflex. Use a retinoscope with a bright halogen light source. Move closer if necessary (remember to adjust for your closer working distance).

- Consider using an auto-refractor. However, you may very well get an inaccurate reading, or the instrument might simply refuse to cooperate with such a difficult eye.

- Put your best estimate of the refraction in the phoropter, including some astigmatic error, and subjectively refine; beginning with large (0.75 D) increments in sphere and a *high power* Jackson cross cylinder. Smaller increments will not produce a discernible change when visual acuity is only 20/400.

- Try a contact lens with over-refraction. Much of the refractive error may be due to irregular astigmatism.
- Consider a stenopeic slit refraction (see following question).
- Do not prescribe "laser pinhole glasses, as seen on TV!" Although they improve visual acuity, they block too much light and impair peripheral vision too much to be recommended.

Describe a stenopeic slit and how to use it in refraction.

A stenopeic slit is an elongated pinhole that may be useful as a guide to determine *subjective* astigmatic refractive error, especially when the retinoscopic reflex is poor (pupil too small, media opacities). Few ophthalmologists use this technique, but demonstrating knowledge of how to use it seems to be a rite of passage in this profession.

Use spheres to put the conoid of Sturm somewhere near the retina. Next, rotate the slit until the patient has the best acuity. If there is no change as the slit is rotated, you may have inadvertently put the circle of least confusion on the retina—change the sphere slightly and try again. Add plus or minus sphere to sharpen the image (position #1). This sphere moves the focal line that is perpendicular to the slit onto the retina. Note the total amount of sphere. Now rotate the slit 90° (position #2) and add plus or minus sphere (you have no way of knowing which it will be) until the sharpest image is obtained. Record the two values on a power cross, and convert to the refraction (see Section 8, *Astigmatism*). This is the refractive correction of one of the principal meridians.

Note that the output of the stenopeic slit is a *power* cross, not an *axis* cross. For example, say that the approximate sphere is +9.00. Stenopeic slit is added and rotated to 165° (position #1). Another +1.00 D is added to obtain best sharpness; total is now +10.00 D. Slit is now rotated to 75° and 4.00 D minus sphere is needed to obtain maximum clarity. The power cross has +10 in the 165° meridian and +6 D acting in the 75° meridian. The corrective lens is therefore +10.00 -4.00 x 165 (or -6.00 +4.00 x 75). This should be refined subjectively, of course.

An important drawback of the stenopeic slit refraction is that the slit must be perfectly centered in the pupil; otherwise the patient will notice little or no change in the image as the slit is rotated. This position can be difficult to maintain, even for a cooperative patient.

You are home for the holidays and your brother-in-law has nearly completed a fascinating 2-hour discourse on all of his previous refractions. He asks whether he is a candidate for "that eye laser." He remembers that a doctor found a "stigma" in his eye. How can you partially evaluate his refractive error simply by inspecting his glasses?

First, you can tell if he is myopic or hyperopic by looking for "with" (myopic) or "against" (hyperopic) movement as you move the glasses back and forth in front of an object. The degree of image magnification (hyperopia) or minification (myopia) allows you to estimate the severity of refractive error. Now look at a round object through the glasses. If it appears oval, the patient has astigmatism. The axis of elongation tells you the axis of minus cylinder. If severe astigmatism is present, it will decrease your enthusiasm for an optimal outcome. Check the lower half of the glasses to look for magnification and peripheral distortion, which indicates that he is wearing progressive addition lenses. Finally, hold the glasses in front of a straight edge (such as a counter top) and check to see if horizontal or vertical prism power is present. Whatever the results of your examination, remember that if you advise him to consider surgery, you will be lucky enough to hear all of the details of surgery and of any imperfect result at every family gathering for the rest of your life.

Caesar Free, a 47-year-old previously emmetropic man with epilepsy, presents to your office complaining of sudden onset of blurry vision and mild eye pain in both eyes. Refraction reveals 4.5 D of myopia in each eye. Why the sudden myopia?

Although diabetes is a strong consideration, other causes must be entertained. Topiramate (Topomax) has been reported to cause sudden myopia and increased intraocular pressure. Two mechanisms for the myopia are likely: 1) forward displacement of the iris-lens diaphragm, and 2) swelling of ciliary body leading to relaxation of zonular tension and subsequent increased convexity of the crystalline lens. Forward displacement of the lens pulls the secondary focal point of the eye into the vitreous (see Chapter 6, *Lens Effectivity and Vertex Distance*). Increased convexity of the crystalline lens also leads to relative myopia due to increased plus power of the refracting elements of the eye. The myopia (and acute glaucoma, which also may occur) will respond to administration of cycloplegic drops.

6

Lens Effectivity and Vertex Distance

What is vertex distance and why is it important?

Strictly speaking, vertex distance is the distance from the front of the cornea to the back of the optical correction. Vertex distance is especially important with higher power spectacle corrections (≥5 D) because lens effectivity changes with distance from the eye in proportion to the power of the lens. It also influences magnification and distortion caused by astigmatic correction.

How do you measure vertex distance?

Vertex distance is most accurately measured with a vertex distance measuring device such as a distometer. The distometer measures the distance from the back of the lens to the surface of the *closed* eyelid. The scale takes the thickness of the eyelid (2 mm) into account—the vertex distance is read directly from the scale on the instrument. It is also possible to simply view the patient in profile, hold a ruler next to the glasses, and estimate the distance. For new prescriptions, a mirror and scale on the side of the phoropter may be used to measure vertex distance. The distance can be adjusted on the phoropter by adjusting the forehead rest using the knob in the center of the instrument.

Why would a hyperopic man slide his glasses down his nose?

Moving a lens, either plus or minus, away from the eye gives it more effective plus power for distance viewing because it pulls its far point with it. Why? Remember that to correct any refractive error with glasses, one must simply find a lens with a focal point that coincides with the far point of the eye (when that lens is held in the spectacle plane). In a hyperope, the far point is behind the eye. If the plus lens is too weak in the spectacle plane (that is, the hyperope is undercorrected), the focal point of the lens is behind the far point of the eye. Moving the lens forward moves the focal point of the lens closer to the far point of the eye (the lens is effectively more powerful). Thus, the hyperope in this example is undercorrected.

In a myope, the far point is in front of the eye. Moving the minus lens forward moves the focal point of the lens further in front of the far point of the eye (the lens again has more plus power, LESS minus power). Why would a myope slide her glasses down her nose if it makes the glasses less powerful? The extra plus power is useful for near work.

Effective plus power is increased for near viewing in all myopes and in high hyperopes; in low hyperopes, the position of the object with respect to the focal length of the lens becomes a contributing factor and effective plus power may or may not increase.

Hunter DG, West CE.
Last Minute Optics: A Concise Review of Optics, Refraction, and Contact Lenses, Second Edition (pp 31-36).
© 2010 SLACK Incorporated

A patient is fully corrected with -10.00 D spectacles at a vertex distance of 10 mm.

A. What correction will be needed if the spectacles are positioned 20 mm away from the eye?

This is a standard vertex distance problem, so draw the eye and the lens and find the far point of the eye. Remember that the patient is a myope, so the far point is *in front of* the eye. The far point is 1 / 10 = 0.1 m = 100 mm in front of the lens, or *110 mm in front of the eye* (it is easiest to convert from m to mm for these calculations) (Figure 6-1).

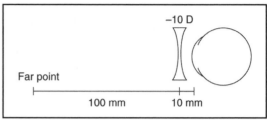

Figure 6-1. Answer to vertex distance problem—first step.

Now that you know the location of the far point, you can forget about where the spectacles used to be (that number will only confuse you). To emphasize this point, we will re-draw the above diagram (Figure 6-2).

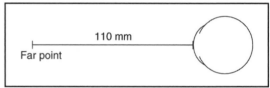

Figure 6-2. Answer to vertex distance problem—second step.

The new spectacle plane 20 mm in front of the eye is (110 – 20) = 90 mm from the far point. The required power is (1 / 0.090 m) = -11.1 D.

B. What if the glasses were moved to a 5 mm vertex distance?

-9.5 D. (110 – 5 = 105; 1 / 0.105 = 9.5).

C. What power soft contact lens will probably be required?

-9.0 D. (1 / 0.110 = 9.1; closest power available is -9.0).

D. What power lens would be needed if it were to be held 10 cm in front of the eye?

-100.0 D. (110 – 100 = 10 mm; 1 / 0.010 = 100).

What if the patient in the previous problem was a +10.00 hyperope? (Answer parts A through D).

In this case, the far point is 1 / 10 = 0.1 m = 100 mm *behind* the lens, or (100 – 10) = 90 mm behind the cornea (Figure 6-3).

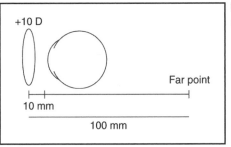

Figure 6-3. Answer to vertex distance problem using +10.0 D lens.

A. +9.1 D. A lens 20 mm in front of the eye is (90 + 20) = 110 mm away from the far point, so the power required is (1 / 0.110 m) = +9.1 D, with 9.0 D the closest power available.

B. +10.5 D. (90 + 5) = 95 mm. (1 / 0.095) = +10.5 D.

C. +11.1 D. (1 / 0.090) = +11.1 D, with 11.0 D the closest power available.

D. +5.25 D. (100 + 90) = 190 mm. (1 / 0.19 m) = +5.25 D. Contrast this with the myopic situation.

You are a -5 D spectacle-corrected myope stranded on a small island with your assistant, Gilligan. Unfortunately, Gilligan has broken your glasses (which had an 11 mm vertex distance). The only lens available to you is a -55 D Hruby lens, which Gilligan was keeping in his footlocker.

A. How many centimeters from the eye should you hold the lens to fully correct your refractive error?

First, locate your far point. The far point is 211 mm in front of the eye (Figure 6-4). The new lens has a power of -55 D, which means its focal point is (1 / 55) = 0.018 m, or 18 mm away from the lens. To correct the refractive error, the focal point of the lens should coincide with the far point of the eye. Therefore it should be 18 mm away from the far point, or (211 – 18) = 193 mm (19.3 cm) in front of the eye.

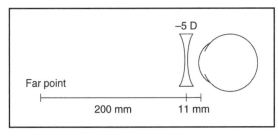

Figure 6-4. Far point of the stranded skipper.

B. Why wouldn't you be able to read the 20/20 line with this correction?

The problem is magnification. This configuration turns the combination of the eye and its corrective lens into a reverse Galilean telescope, where the eyepiece is approximately +5 D (the "error lens" of the myopic eye) and the objective is -55 D. The resulting magnification is

(5/55) ≅0.1X. Thus, the 20/20 line, while in focus, subtends 1/10 of the angle it would in the eye of an emmetrope. The best distance visual acuity attainable is only about 20/200 (assuming an otherwise normal eye). Similarly, properly corrected patients with very high myopia may not be able to read 20/20 through their spectacle lenses even in the absence of other pathology.

This magnification effect gives refractive surgeons an advantage, for moving the point of correction of a myopic patient from the spectacle plane to the corneal plane automatically increases the magnification of the image, such that best-corrected visual acuity should actually improve in these patients. Conversely, the best-corrected visual acuity in very high hyperopes may decline slightly when the refractive correction is moved from the spectacle plane to the corneal plane.

7

Accommodation, Presbyopia, and Bifocals

What is accommodation? How do you describe a patient's ability to accommodate?

The eye can increase its total dioptric power by increasing the convexity of the lens through ciliary muscle contraction. For simplicity, assume that this additional plus power is acting at the cornea.

- *Near point* is the point of clearest vision (conjugate to the retina) when accommodation is maximally active. Note that this is a location measured from the cornea in centimeters or meters. (Compare with *far point*, the point that is conjugate to the retina when accommodation is fully relaxed. A myopic eye has a far point between infinity and the cornea, while a hyperopic eye has a far point somewhere behind the eye.)
- *Amplitude of accommodation* is the maximum number of diopters that the eye can accommodate. Note that this is expressed in diopters.
- *Range of accommodation* is the linear distance over which a patient can accommodate and maintain clear vision. Note that this is expressed as two locations measured in centimeters or meters from the eye.

How do you measure accommodative amplitude?

Accommodative amplitude is quantified by asking the patient to maximally accommodate on a near target. To perform this measurement properly, one eye is occluded. The best distance correction should be in place. Slowly move a target closer to the eye while strongly encouraging the patient to keep it in focus. The target is normally slid along a Prince rule or an RAF (Royal Air Force) rule, which are rulers scaled in both diopters and centimeters. Record the point at which the patient is no longer able to keep the target in focus. Repeat for the other eye.

To perform the test in patients with poor accommodation, you must begin the test with a +3.00 add to bring the far point to a more practical distance for testing. If you use a +3.00 add, remember to subtract 3 D from the accommodative amplitude you measure.

How much accommodation is normal? Should bifocals be prescribed according to tables?

According to published tables, a 40-year-old has about 6 D, a 44-year-old 4 D, and a 60-year-old about 1 D of accommodation. There are tables and formulas to guide the refractionist, but they are generally just a starting point.

The formal procedure for prescribing bifocals requires determination of the patient's needs and preferred working distances. This gives the total amount of accommodation plus

Hunter DG, West CE.
Last Minute Optics: A Concise Review of Optics, Refraction, and Contact Lenses, Second Edition (pp 37-44).
© 2010 SLACK Incorporated

add that will be required, for example, +3.00 D if the preferred working distance is 33 cm. Now measure the amplitude of accommodation in each eye. Let the patient use half of his accommodative amplitude while reading, and put the rest in a bifocal.

Realistically, we don't have time to measure the amplitude of accommodation in every presbyope, but this procedure should be considered for any special cases. For example, measure accommodative amplitude if there is a question about whether bifocals are indicated (eg, a 35-year-old with asthenopia), if a patient is not happy with new reading glasses, or if there is concern about accommodative insufficiency from drugs, trauma, or other disease. If these don't apply, it is acceptable to estimate accommodative amplitude when selecting the add, as long as you take the time to place it in front of the distance correction while the patient views familiar reading material (such as a newspaper). Remember to take into account other needs, such as viewing a computer monitor at intermediate distances.

At 5:30 PM in the middle of a busy Friday clinic, you receive a phone call from the mother of an 18-month-old child that you examined this morning. She says that the child has developed a fever and seems quite fussy. Her face seems red. What do you think might be happening, and how should you proceed?

The child may have developed atropine toxicity, either from drops you administered or because she ingested the bottle of atropine drops that you dispensed. This is a toxic reaction (an anticholinergic overdose), not an idiosyncratic or allergic reaction, though some children are more sensitive than others. Other symptoms and signs include dry mouth, rapid pulse, and nausea. The key finding in this case is that the child is "fussy." Since it is possible that this is a severe reaction, you need to bring the patient in or refer her to a nearby emergency room for prompt evaluation. If there is a significant overdose, supportive therapy (ie, sponge baths for fever, catheterization for urinary retention, oxygen for mild respiratory depression) is usually enough, but severe reactions may require treatment with intravenous physostigmine or intubation for respiratory depression. The initial dosage of physostigmine is 0.5 mg for children or 2 mg for adolescents. In cases where significant ingestion of atropine has occurred, induce emesis and give oral pilocarpine (5 mg repeated until mouth is moist).

What are some other side effects of cycloplegic agents?

Any of the cycloplegic agents can cause mental status changes including hallucinations, ataxia, incoherent speech, restlessness, hallucinations, hyperactivity, disorientation, or seizures. Systemic consequences (described previously) are often the first to be observed in small infants. Some authorities recommended that infants not be fed for 4 hours after cycloplegic agents are given due to feeding intolerance that may develop. Scopolamine and higher percentages of cyclopentolate (2%) and homatropine (5%) are most commonly associated with central nervous system manifestations. Remember to adjust the dosage accordingly, using the lowest percentage of cycloplegic agent that you need. For example, in neonates, the commercial formulation "CycloMydril" contains very weak concentrations of cyclopentolate (0.2%) and phenylephrine (1%). Toddlers, preschoolers, and those with darkly pigmented irides will require the stronger, more concentrated cycloplegic agents. Remember, phenylephrine is often used along with other agents, but it has no cycloplegic effect—it is a sympathomimetic agent used for pupillary dilation alone. You should be able to describe the rationale for the cycloplegic regimen that you use in your patients.

What are four types of hyperopia?

1. *Absolute hyperopia.* Without cycloplegia, this is the least amount of plus correction required for clear vision at distance.

2. *Manifest hyperopia.* Without cycloplegia, this is the most plus correction the eye can accept without blurring of vision.

3. *Facultative hyperopia.* This is the difference between absolute and manifest hyperopia.

4. *Latent hyperopia.* This is the difference between manifest hyperopia and hyperopia measured with cycloplegia.

A patient requires +1.00 D to see at distance. Manifest refraction reveals she will tolerate up to +2.00 D. Cycloplegic refraction is +5 D sphere. What are the absolute, manifest, cycloplegic, facultative, and latent hyperopia?

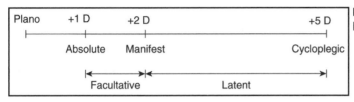

Figure 7-1. Types of hyperopia.

This is easiest to picture by considering a diagram (Figure 7-1).

- Absolute = +1 D
- Manifest = +2 D
- Cycloplegic = +5 D
- Facultative = 2 − 1 = +1 D
- Latent = 5 − 2 = +3 D

An emmetropic child named Lisa Confusion has an amplitude of accommodation of 50 D.

A. What is Lisa's range of accommodation?

Infinity to 2 cm. Range of accommodation is specified in terms of distance. Without accommodation, she is in focus at infinity—her far point. With maximal accommodation, she is in focus at (1 / 50) = 0.02 m = 2 cm—her near point (Figure 7-2).

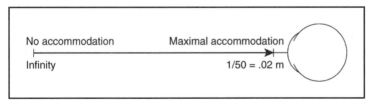

Figure 7-2. Lisa's range of accommodation.

B. The child's older cousin, Prentice S. Rewel, is a +8 D hyperope with a 20 D amplitude of accommodation. What is his near point and range of accommodation when he is not wearing spectacles?

Prentice has to "use up" 8 D of his accommodative amplitude just to be in focus at infinity. He uses the remaining (20 – 8) = 12 D for near vision. Therefore his near point is (1 / 12) = 0.083 m from his cornea, *not* (1 / 20). His range of accommodation is from infinity to 8.3 cm (Figure 7-3).

Figure 7-3. Prentice's range of accommodation when he is not wearing glasses.

Accommodating 8 D · Infinity · Accommodating 20 D · 1/(20–8) = .083 m

C. What is Prentice's near point and range of accommodation when he is wearing his +8.00 D spectacles?

In this case, Prentice does not have to "use up" any of his accommodative amplitude, so he can apply the full force of his 20 D to his near vision. The result is a near point of 1 / 20 = 0.05 m. The range of accommodation is infinity to 5 cm.

Note that this is another demonstration of the non-linearity of diopters. That is, the additional 8 D of available accommodation only added 3.3 cm to his range of accommodation.

D. The child's mother, Lady Frenella Prism, is a 15 D myope with a 10 D amplitude of accommodation. What is her range of accommodation with and without spectacles?

With spectacles: Infinity to 10 cm. Without spectacles: 6.7 cm to 4.0 cm.

With spectacles, we know that Frenella is in focus at infinity when accommodation is relaxed. When she maximally accommodates, she is in focus at (1 / 10) = 0.1 m = 10 cm in front of the eye.

Without spectacles, things are very different with the myope than with the hyperope (Figure 7-4). Uncorrected myopes have a far point that is not all the way out at infinity—think of it as a "head start" on accommodation. In Frenella's case, her uncorrected 15 D of myopia are like 15 D of built-in accommodation. Thus, her farthest point of clear vision without correction is (1 / 15) = 6.7 cm in front of the eye. When she does accommodate, the 10 D that she can generate adds to the 15 D that is already built in, and her near point is (1 / 25) = 0.040 m = 4 cm.

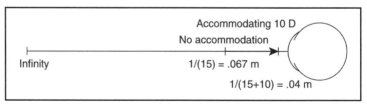

Figure 7-4. Lady's range of accommodation without spectacles.

Accommodating 10 D · No accommodation · Infinity · 1/(15) = .067 m · 1/(15+10) = .04 m

E. The child's great grandfather, Conrad F. Sturm, has a range of accommodation of 20 cm (distance) to 15 cm (near) without spectacles. What is his distance refractive correction?

When accommodation is relaxed, Mr. Sturm is in focus at 20 cm (his far point). The reciprocal of his far point is (1 / 0.2 m) = 5 D. Ignoring vertex distance, his distance correction is therefore -5 D (Figure 7-5).

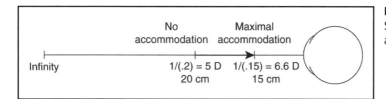

Figure 7-5. Mr. Sturm's range of accommodation.

F. What is Mr. Sturm's amplitude of accommodation?

To see clearly at 15 cm, an emmetrope requires (1 / 0.15) = 6.6 D of accommodation. However, Mr. Sturm already has a 5 D "head start" on accommodation when he is not wearing spectacles. Therefore he must only accommodate (6.6 – 5.0) = 1.6 D to see clearly at 15 cm. This is all he is capable of, so his amplitude of accommodation is 1.6 D.

G. What is Mr. Sturm's range of accommodation with spectacles?

Infinity to 62 cm. When he wears his -5 D refractive correction, he is in focus at infinity with accommodation relaxed. When he uses his full 1.6 D of accommodative amplitude, he can see clearly at (1 / 1.6) = 0.62 m = 62 cm. He is a good candidate for a reading add.

Name some causes of inadequate accommodation.

- *Lens changes:* normal presbyopia
- *Severe refractive error:* latent hyperopia
- Inadequate accommodation
  - Systemic factors
    - Oral medications (parasympatholytics, phenothiazines, tranquilizers, chloroquine)
    - Down syndrome
    - Concurrent systemic illnesses (hypothyroidism, severe anemia, myasthenia gravis, diabetes)
    - Prior encephalitis or meningitis
  - Remote factors
    - Tumor
    - Head trauma
  - Local factors
    - Tonic pupil
    - Cycloplegic use
    - Ocular trauma

A 21-year-old college junior presents to your clinic 1 week before his final exams with a new onset intermittent esotropia associated with poor distance vision. What is your diagnosis and what would you do to help confirm it? How about treatment?

Spasm of accommodation; look for small pupils when he is esotropic. Explain the problem to the patient and try to alleviate his fears. Ask him to look away from his near work frequently to relax accommodation, and/or offer reading glasses/bifocals. Cycloplegia can also be used to break accommodative spasm.

Discuss the management of an early hyperopic presbyope who has never worn glasses.

Perform a manifest and cycloplegic refraction, followed by a post-cycloplegic trial of your findings. Listen to his needs and concerns about working distance, type of work performed, and cosmesis. Discuss single vision versus bifocals for correction; often single vision glasses, correcting only the manifest hyperopia, will be adequate (for a while!). If bifocals are given, warn the patient that a change in glasses may be necessary again in the near future, as he learns to relax and use his add.

A 41-year-old myopic man (who is recently divorced and sporting a new goatee and impressive gold jewelry) is interested in getting contact lenses for the first time. After a careful cycloplegic refraction, your technician dispenses -10.00 D well-fitting, daily wear soft lenses that allow 20/15 vision at distance. Complete care instructions are given and understood. He returns a few days later, absolutely livid. What is he so angry about?

He can no longer see up close. This should lead to a discussion of increased accommodative demands (for near work) when a myope's corrective lens is moved closer to the eye. The change in accommodative demand can be calculated, but knowledge of the end result is more important than the mathematical gymnastics required to get there.

Even without performing math, it is possible to understand why a patient must accommodate different amounts when wearing contact lenses versus spectacles. Light rays from a distant object are parallel; therefore, the vergence of light is zero when it reaches the corrective lens, no matter whether the patient is wearing glasses or contact lenses. Light rays from a near target are not parallel—they are divergent. To supply additional plus power required to view a near target, the patient must accommodate to compensate for this additional divergence. But how much vergence is present at the cornea? It depends on whether glasses or contact lenses are worn. The vergence of light from a near object is different when it reaches the spectacle plane than it is when it reaches the front surface of the contact lens. It turns out that after considering vertex distance and vergence, light is more divergent after passing through a minus contact lens than it is by the time it reaches the cornea after passing through the appropriate spectacle lens. Thus, the myope must accommodate more through a contact lens than through a spectacle lens. In contrast, a hyperope has more trouble accommodating through glasses than through contact lenses.

A quick way to remember that the myope will have more trouble is to consider that when the myope is reading, his eyes are converged such that the line of sight passes nasal to the optical centers of the spectacle lenses. This provides a bit of effective base-in prism, which aids convergence (see the Prisms section). When the glasses are removed, the convergence aid is lost, and the myope will have more trouble reading with contact lenses. The opposite is true with hyperopes. This effect is less important than the accommodative effect, but easier to figure out.

What are progressive addition lenses, and how might you counsel a potential progressive addition lens wearer?

Progressive addition lenses (PALs) have a channel of progressively higher (plus) dioptric power on the front of the lens, effectively creating a variable focus bifocal. While there is no line to bother the cosmetically conscious, there is marked irregular peripheral astigmatism. This can be quite annoying, especially in the higher power PALs. Some opticians seem to have difficulty adjusting the "segment height" of a progressive lens, especially when patients choose frames that are prone to misalignment. Patients who have already worn traditional bifocals have the greatest difficulty accepting the distortion of a progressive addition lens. Newly diagnosed presbyopes requiring a low power add are most likely to do well with progressive addition lenses. There are many different brands and configurations of progressive lenses. "Hard" design PALs have a wider near zone at the expense of more peripheral aberrations, while "soft" design PALs have a narrower near reading area with lesser amounts of peripheral aberrations. If the patient is not satisfied with the first choice, suggest that he or she consider trying another (possibly more expensive) design.

8

Astigmatism

What is the conoid of Sturm?

The conoid of Sturm is the three-dimensional envelope of light rays formed by an astigmatic lens acting upon the rays of light from a point object. The conoid of Sturm contains the perimeter of the lens, the two focal lines, and the circle of least confusion. An astigmatic (*a-stigmatic*, no point) eye has two far lines rather than a single far point.

What is the circle of least confusion?

When light passes through a spherocylindrical (astigmatic) lens, it is focused not to a point, but into two successive focal lines. Each focal line is formed by the power of the lens acting in the meridian that is at right angles to the focal line. The average dioptric power of the lens, called the spherical equivalent of the lens, is associated with a plane halfway between the two focal lines (in diopters, NOT distance). In this plane, the light is equally blurred in all meridians. Letters are theoretically easiest to recognize at this point, for the blur is equal in all meridians. This location is called the circle of least confusion; it is a circle because its perimeter is defined by the round pupil aperture.

What types of ametropia are represented in Figure 8-1?

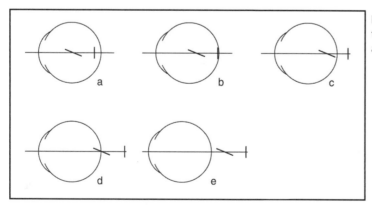

Figure 8-1. Name the different types of ametropia.

The figures represent the focal *lines* of an eye: a. compound myopic, b. simple myopic, c. mixed, d. simple hyperopic, and e. compound hyperopic astigmatism.

Hunter DG, West CE.
Last Minute Optics: A Concise Review of Optics, Refraction, and Contact Lenses, Second Edition (pp 45-56).

What does "with-the-rule" astigmatism mean?

With-the-rule astigmatism is corrected with plus cylinder at 90° or minus cylinder at 180° (ie, the vertical meridian is steeper). The axis doesn't have to be *exactly* 90° or 180°; about 20° on either side qualifies. Outside of those limits, the astigmatism becomes oblique. With- and against-the-rule conventions allow plus and minus cylinder aficionados to converse without translating plus to minus cylinder and vice versa.

Children tend to have with-the-rule astigmatism. This may be due to the elasticity of the eyelids in children, or the pliability of the cornea. The tight lids press down on the upper and lower cornea and steepen its vertical meridian. Older adults have flabby, stretched-out eyelids and thus they tend to have against-the-rule astigmatism… at least until a cataract surgeon ties the 12 o'clock suture too tight and converts the astigmatism back to with-the-rule!

During a subjective refraction, you increase the cylinder 0.50 D. How much should you change the sphere and in which direction? Why? Answer for both plus and minus cylinder.

You should change the sphere 0.25 D in the opposite direction to keep the circle of least confusion on the retina. If you are refracting in plus cylinder and you add +0.50 cylinder, you should change the sphere by -0.25 D. Similarly, if you are refracting in minus cylinder and you add -0.50 cylinder, you should change the sphere by +0.25 D.

What is the spherical equivalent of -1.00 +2.00 x 045?

Plano. Add one-half the cylinder to the sphere; be sure to keep track of minus and plus. Note that this happens to describe a ±1 D Jackson cross cylinder (Figure 8-2). (Note it is *not* a ±2 D Jackson cross cylinder.)

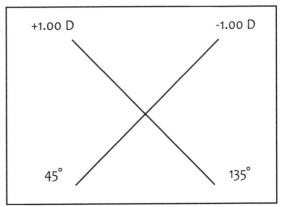

Figure 8-2. A ±1 D Jackson cross cylinder. Rx = -1.00 + 2.00 x 045. See below regarding the value of the axis.

+1.00 D   -1.00 D

45°   135°

Write a prescription for a ±0.50 D Jackson cross cylinder in both minus and plus cylinder notation.

+0.50 -1.00 x 090 (minus cylinder)
0.50 +1.00 x 180 (plus cylinder)

Note that the axis is specified at random (it depends on how you are holding the cylinder); as long as the axis of the minus cylinder form is 90° away from that of the plus cylinder form, you have the correct answer.

Why are Jackson cross cylinders available with a variety of powers?

Patients with poorer visual acuity need to be shown a bigger difference for comparison when subjectively refining cylinder axis and power. As a general guide, use ±0.12 D for 20/15 to 20/20, ±0.25 D for 20/25 to 20/30 (this is built into a typical phoropter), ±0.50 D for 20/40 to 20/60, and ±1.00 D for 20/70 to 20/200.

What is the spherocylindrical notation, both plus and minus, for the combination of the following 2 cylindrical lenses (note this is axis notation):

Lens #1: +3.00 x 170

Lens #2: -5.25 x 080

+ 3.00 -8.25 x 080   *or*   - 5.25 +8.25 x 170

There are a number of different approaches to this type of a problem. One way may seem more reasonable to you than another depending on your problem-solving style. Three approaches will be presented here.

• Method 1

1. Draw the 2 lenses as a power cross (Figure 8-3).

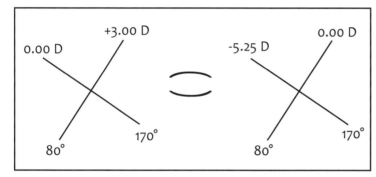

Figure 8-3. Two lenses drawn as a combination of 2 power crosses. The two curved lines indicate "combined with".

2. Add +3.00 to the horizontal arm on the left. To avoid changing combined power of the 2 power crosses, you must then subtract +3.00 from the horizontal arm on the cross on the right (Figure 8-4).

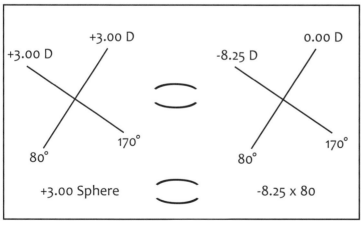

Figure 8-4. Conversion of power cross to spherocylindrical notation is completed by adding equal and opposite powers to corresponding meridians.

3. Now you simply copy the prescription from directly from the power crosses:

+3.00 -8.25 x 80 (note the axis).

Had you started by adding -5.25 D to the vertical arm of the power cross on the right, it still would have come out correctly (in plus cylinder) as long as you remembered to subtract -5.25 D from the vertical arm of the cross on the left (+3.00 − (-5.25) = +8.25). Rx = -5.25 +8.25 x 170.

• Method 2

Here is another 3-step method that involves more memorization but requires less understanding.

1. Select 1 lens and declare that it will be the spherical part of the spherocylindrical lens. For minus cylinder notation, select the more positive or less negative lens.

2. Find the cylindrical part, which is the difference in power between the 2 lenses.

3. Find the axis of the correcting cylinder.

In the question above, make lens #1 the sphere (sphere = +3.00). Now find the difference (lens #2 − lens #1 = -8.25 D). The axis of lens #2 is the axis of the correcting cylinder (80°). Be careful not to subtract the wrong lenses, or you'll get the wrong sign on the cylinder power.

• Method 3

Add equal and opposite (canceling) cylinders to the meridional powers. That is, you can add a combination of a +3.00 and a -3.00 D lens in the same meridian to the 2 lenses above without changing the overall power of the combination:

$$+3.00 \text{ x } 170 \supset -5.25 \text{ x } 080$$
$$+3.00 \text{ x } 080 \supset -3.00 \text{ x } 080$$
$$\overline{\phantom{+3.00 \text{ x } 080 \supset -3.00 \text{ x } 080}}$$
$$+3.00 \text{ sph} \quad \supset -8.25 \text{ x } 080$$

Choose whichever method makes the most sense to you, and then apply it consistently. It is important to practice. To check your work, draw a power cross from the prescription above, and it should bring you back to the original combination of 2 cylindrical lenses.

Convert the power crosses in Figure 8-5 to spectacle corrections. Give both plus and minus cylinder notations.

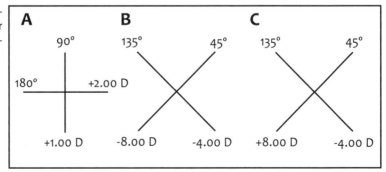

Figure 8-5. Convert the power crosses to spectacle corrections.

Remember that a cylindrical lens held at a given *axis* provides *power* in a meridian 90° away from the axis. Some people prefer to work strictly with power crosses, remembering to add or subtract 90° when writing the spectacle prescription indicated by the power cross. Others find that they are less likely to confuse meridians by converting the power cross to an axis cross before making any other calculations. If you choose to use an axis cross, **label carefully!** We will proceed directly from the power cross for this set of problems.

A. Plus cylinder form: +1.00 +1.00 x 90
   Minus cylinder form: +2.00 -1.00 x 180

The approach to this problem is similar to that of the previous problem. Use any of the 3 methods. Or, here is another modification of the second approach: For plus cylinder notation, start with the meridian having the least plus power (or most minus): +1.00. Now ask, how many diopters of plus cylinder must be added to obtain the power in the other meridian? (+1.00 D) What axis specifies the meridian requiring this additional power? (90°) Thus, the spectacle correction is +1.00 +1.00 x 90.

To double check, convert from the power cross directly into minus cylinder form, then compare the plus cylinder to the minus cylinder form and see if they match. Or convert back to a power cross and see if you get what you started with.

B. Plus cylinder: -8.00 +4.00 x 45
   Minus cylinder: -4.00 -4.00 x 135

(Least plus [most minus]: -8.00. Added power: +4.00. Axis of added power: 45. OR most plus [least minus]: -4.00. Added power: -4.00. Axis of added power: 135)

C. Plus cylinder: -4.00 +12.00 x 135
   Minus cylinder: +8.00 -12.00 x 45

Note that to go from -4 D to +8 D, a cylindrical lens of 12 D is required.

For each of the previous lenses, calculate the spherical equivalent. How far are the 2 focal lines and the circle of least confusion from the lens, assuming a point object arising at infinity?

A. Spherical equivalent +1.50 D   Circle of least confusion: +0.67 m

   Focal line of power at 180: +1 m   Focal line of power at 90: +0.5 m

B. Spherical equivalent -6.00 D   Circle of least confusion: -0.167 m

   Focal line of power at 135: -0.125m   Focal line of power at 45: -0.25 m

C. Spherical equivalent +2.00 D   Circle of least confusion: +0.5 m

   Focal line of power at 135: +0.125m   Focal line of power at 45: -0.25 m

For part A, the spherical equivalent = (sphere) + (1 / 2) x (cylinder) = (+1.00) + (1 / 2) x (+1.00) = +1.50. The circle of least confusion is at the focal plane of the spherical equivalent (1 / +1.50) = +0.67 m. The plus sign indicates that it is to the right of the lens. The focal line in the 180° meridian is formed by the power acting 90° away, in the 90° meridian. Since the power in the 90° meridian is +1.00 D, the focal line in the 180° meridian will be formed 1 meter to the right of the lens. Similarly, the focal line in the 90° meridian is (1 / 2) = 0.5 m away.

For part B, the focal lines and circle of least confusion are all to the left of the lens. For part C, the focal lines are on either side of the lens.

Convert the following prescriptions to plus or minus cylinder notation. Compute the spherical equivalent and draw the power cross. What type of astigmatism is present? Is it with-the-rule, or against-the-rule, or oblique?

    A. +3.00 -2.00 x 80

    B. +1.00 -4.00 x 80

    C. -5.00 +9.00 x 90

A. +3.00 -2.00 x 80. Compound hyperopic, against-the-rule.

Plus cylinder: +1.00 +2.00 x 170. Spherical equivalent +2.00 D. Now for a little trick: take the *spherical* power of the lens in plus cylinder format (in this case, +1.00) and place it on the power cross in the meridian specified by the axis of the *cylindrical* power in the plus cylinder format (in this case, 170°). Do the same thing with the spherical power in the minus cylinder format (in this case, +3.00, 80°). The result is the correct power cross (Figure 8-6)!

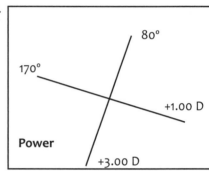

Figure 8-6. Power cross for +3.00 - 2.00 x 80.

Now another trick. Look at the spherical power in plus cylinder format. It is positive. Now look at the spherical power in minus cylinder format. It is also positive. Since both are positive, this is compound hyperopic astigmatism. See Table 8-1 for other situations.

Table 8-1

Rapid Way to Determine Type of Astigmatism

MINUS CYLINDER FORMAT (SPHERICAL POWER)	PLUS CYLINDER FORMAT (SPHERICAL POWER)	TYPE OF ASTIGMATISM
Positive	Positive	Compound hyperopic
Negative	Negative	Compound myopic
Plano	Negative	Simple myopic
Positive	Plano	Simple hyperopic
Negative	Positive	Mixed
Positive	Negative	Mixed

B. +1.00 -4.00 x 80. Plus cylinder: -3.00 +4.00 x 170. Spherical equivalent -1.00 D; mixed, against-the-rule (Figure 8-7).

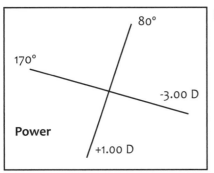

Figure 8-7. Power cross for +1.00 - 4.00 x 80.

C. -5.00 +9.00 x 90. Minus cylinder: +4.00 -9.00 x 180. Spherical equivalent -0.50 D; mixed, with-the-rule (Figure 8-8).

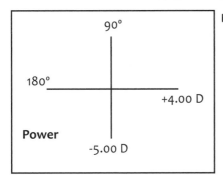

Figure 8-8. Power cross for -5.00 + 9.00 x 90.

You are performing retinoscopy on a child using free spherical lenses. When you orient the streak in the horizontal meridian and sweep vertically, you neutralize the reflex with a +3.00 D lens. When you orient the streak in the vertical meridian and sweep horizontally, you neutralize the reflex with a +4.00 D lens. What glasses prescription would you prescribe to fully correct this refractive error?

+1.50 +1.00 x 90. The retinoscopy results are easiest to summarize with an *axis* cross. First, you swept the horizontal streak up (Figure 8-9).

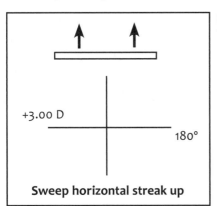

Figure 8-9. Retinoscopy: Streak oriented in horizontal meridian, sweeping vertically.

Now sweep the vertical streak across (Figure 8-10).

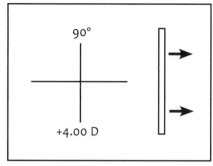

Figure 8-10. Retinoscopy: Streak oriented in vertical meridian, sweeping horizontally.

This translates to a spherocylindrical correction of +3.00 +1.00 x 90 (remember, you drew an axis cross so you don't change the cylinder axis when converting from the cross to a glasses prescription). Don't forget to subtract your working distance! If it is 0.67 m, the refraction is +1.50 +1.00 x 90.

A spherocylindrical lens converts light from a point object at infinity to a horizontal line at +0.33 m and a vertical line at +0.5 m. Draw the power cross for the lens. With light traveling from left to right through the lens, where is the circle of least confusion for the conoid of Sturm thus formed? What is the spherocylindrical notation for this lens?

40 cm to the right of the lens. +2.00 +1.00 x 180, or +3.00 -1.00 x 90 (Figure 8-11).

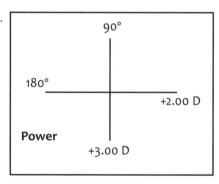

Figure 8-11. Power cross for +2.00 + 1.00 x 180.

The focal line is parallel to the axis, perpendicular to the power. The horizontal line at +.33 m is due to a vertically oriented cylindrical lens of (1 / 0.33) = +3 D, axis 180. The vertical line (90°) is due to a horizontally oriented cylindrical lens power of (1 / 0.5) = +2 D, axis 90. Therefore the power cross will have 3.00 D acting in the 90° meridian and +2.00 D acting in the 180° meridian. The spherical equivalent is thus ([+2] + [1 / 2] [1]) = +2.50. The circle of least confusion is located (1 / +2.50) = +0.40 m, or 40 cm to the right of the lens.

Sixteen weeks after extracapsular cataract extraction using the nuclear expression technique, the following keratometry readings are obtained:

40.00 @ 30°

43.00 @ 120°

A. What power and axis of cylindrical correction is required to correct the postoperative astigmatism, assuming all astigmatism is corneal? Assuming a single, tight, radially placed suture is present, in what meridian would you be most likely to find it? (For extra credit: What is the glasses prescription if you know that the spherical equivalent is +1.00?)

There is a 3 D difference between meridians. Ignoring vertex distance, this is corrected with a 3 D cylindrical spectacle lens. The 120° meridian is the steepest; thus, it has the most plus power. It would be neutralized by -3.00 D of power in that meridian. But astigmatic refractive power exerted in the 120° meridian has its axis 90° away (at 30°). Thus, the glasses should have -3.00 D x 30 (or +3.00 x 120).

Extra credit: Ignoring vertex distance effectivity changes, a plano spherical equivalent of the required cylindrical correction is +1.50 -3.00 x 30, or -1.50 +3.00 x 120. For a spherical equivalent of +1.00, the prescription must be +2.50 -3.00 x 30, or -0.50 +3.00 x 120.

B. A tight suture steepens the cornea. Why is this so, especially given that the cornea is flattened in the area immediately under the suture?

Recall that the circumference of the globe is constant. Therefore, flattening in one area has to be balanced by steepening elsewhere in the globe. This compensatory steepening takes place in the center of the cornea overlying the pupil (Figure 8-12). Try this demonstration: Hold an index card horizontally, giving it a gentle convex curvature. Now without moving the edges of the card, flatten a small section near one edge (this is the tight suture). Note how the rest of the card becomes steeper. In this example, the tight suture (steep meridian) is at 120° (11 o'clock). Therefore, this suture should be cut.

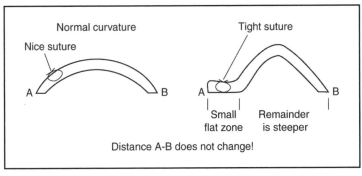

Figure 8-12. How a tight suture steepens the cornea.

A 23-year-old man returns to your office 1 hour after receiving the spectacles you prescribed for him. He had never worn glasses before. He says the whole world is like a fishbowl, he feels that the floor is coming up at him, doors are falling toward him, he is dizzy and has a headache, and his girlfriend just broke up with him. Visual acuity is 20/50 OU without correction and 20/20 OU with correction.

A. What's wrong?

The prescription most likely called for oblique cylinder in one or both eyes. Correction of oblique astigmatism can cause fairly subtle distortion under monocular viewing conditions, with tremendous distortion of depth under binocular conditions. The easiest example to consider is a door frame. If astigmatic correction is with- or against-the-rule, the door frame may look elongated or shortened but the left side will be just as tall as the right side. With correction of oblique astigmatism, the right eye may see the door frame tilted slightly to the left, which may or may not be noticeable. However, if the left eye sees the door frame tilted slightly to the right, then the binocular visual system (which is sensitive to extremely small disparities between the two eyes) kicks in (Figure 8-13). The brain will think that the top of the door is closer to the patient than the bottom. He will feel that it is falling toward him. Another possibility is that there is anisometropia with aniseikonia.

Figure 8-13. Depth distortion resulting from astigmatism. Left eye sees image on the left; right eye sees image on the right. Result is a stereogram.

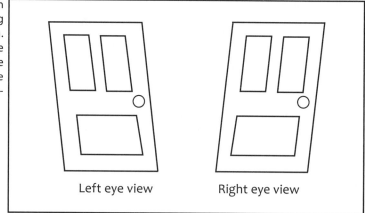

Left eye view

Right eye view

B. How can this be helped?

- Reassure patient that he will get used to it (the younger the patient and the lower the astigmatism, the more likely this will be true)
- Minimize vertex distance
- Be sure lenses are of minus cylinder type (cylinder ground on the back surface)
- Rotate cylinder toward the 90° or 180° meridian (will compromise visual acuity)
- Decrease power of cylinder (will compromise visual acuity)
- Consider contact lenses (no compromise of visual acuity)

Aberrations, Distortions, and Irregularities

What is spherical aberration? Please give some examples.

Rays that strike the periphery of a lens system will be more strongly bent (if "positive" spherical aberration is present), effectively shortening its focal length. This becomes a problem with widely dilated pupils (pharmacologic or in low light levels), causing the patient to become more myopic.

An ophthalmology resident is performing retinoscopy on a cooperative, cyclopleged 5-year-old child when she notices that the center of the pupil shows a "with" reflex, while the periphery shows "against" movement.

A. What optical phenomenon is responsible?

Spherical aberration. In a biconvex lens such as the human lens, the peripheral rays are refracted more strongly than the central rays. The peripheral lens thus appears more myopic during retinoscopy than the center, especially when the pupil is widely dilated (as in children).

B. Which reflex should the resident neutralize?

She should neutralize the central part, which will be most relevant to the refraction once the drops wear off and the pupil returns to its normal size.

A senior ophthalmology resident performs retinoscopy on an uncooperative 4-year-old boy, and records OD +4.50 +1.00 x 180, OS +4.50 sphere. The attending ophthalmologist repeats the retinoscopy and finds +4.50 sphere OU. Assuming the attending's retinoscopy is correct, what happened?

During the resident's retinoscopy of the right eye, the child was trying to spit on the resident. She moved to the patient's right and continued the retinoscopy while the boy was distracted by a toy slightly to his left. This caused astigmatism of oblique incidence, with plus cylinder induced in the axis of tilt of the boy's eye (90°), and was corrected by additional plus cylinder at 180°. The attending performed retinoscopy on axis.

Hunter DG, West CE.
Last Minute Optics: A Concise Review of Optics, Refraction, and Contact Lenses, Second Edition (pp 57-60).
© 2010 SLACK Incorporated

What is astigmatism of oblique incidence? Pantoscopic tilt? Why would a patient tilt his glasses?

Tilting a spherical lens adds both sphere and cylinder of the same sign of the original lens, with the cylinder axis in the axis of the tilt. This is "astigmatism of oblique incidence," or "radial astigmatism." Tilting a -10.00 D lens 10° forward results in -10.10 -0.31 x 180; tilting a +10.00 lens forward 20° results in +10.41 +1.38 x 180. It is not important to remember the *exact* amounts of sphere and cylinder induced—just remember that pantoscopic tilt changes both sphere and cylinder power induced.

The curvature and tilt of spectacle lenses is designed to minimize astigmatism of oblique incidence. Pantoscopic tilt is the forward tilt of the spectacle plane relative to vertical. If spectacle lenses were perpendicular to the horizon, there would be significant astigmatism of oblique incidence in the reading position. If they were optimized for reading position, astigmatism of oblique incidence would be maximal for distance viewing. Pantoscopic tilt is the slight (7°) forward tilt of a spectacle lens that is a compromise between the optimal position of the lenses for distance and near work.

People with undercorrected myopia often tilt their glasses to gain more minus sphere. People with undercorrected hyperopia could do the same, but they usually slide their glasses down their nose to gain more plus power instead.

What is coma?

It is the off-axis effect of spherical aberration. Coma causes light rays to be distributed in a pattern similar to that of a comet. Coma generally increases as an object moves away from the optical axis.

10

Contact Lenses

Describe how you would fit a patient with contact lenses.

1. Obtain an accurate refraction.

2. If considering a rigid lens, convert to minus cylinder form and drop the minus cylinder (use spherical power only; the tears form a cylindrical "lens" between the cornea and the contact lens that will correct corneal astigmatism).

3. Calculate correction for zero vertex distance.

4. Evaluate the anterior segment with a slit lamp exam
   - Look for corneal edema, vascularization, staining; note contour.
   - Note tear film and the tear film breakup time (normally >15 seconds). Consider Schirmer testing.
   - Look for lid abnormalities, flip lid to check for papillae, note palpebral fissure.
   - Check for eccentric pupil.

5. Perform keratometry; compare results with spectacle refraction to detect lenticular astigmatism.

6. Discuss rigid vs. soft lenses with patient.

7. Fit lenses (see later discussion)
   - Rigid lens: Evaluate fluorescein pattern.
   - Soft lens: Steep lenses don't move with a blink; flat lenses move too much.

A contact lens is labeled 8.9/13.5/+12.50. What do the numbers mean?

The base curve is 8.9 mm, the diameter of the lens is 13.5 mm, and the power is +12.50 D.

How many diopters of power correspond to a corneal radius of curvature of 8.9 mm?

Using the standard keratometric formula (which is derived in Section 17, *Instruments*):

$$D_{cornea} = \frac{0.3375}{r \text{ (meters)}} = \frac{337.5}{r \text{ (mm)}} = \frac{337.5}{8.9} = 37.9 \text{ D}$$

Hunter DG, West CE.
Last Minute Optics: A Concise Review of Optics, Refraction, and Contact Lenses, Second Edition (pp 61-64).
© 2010 SLACK Incorporated

Note that this is an approximation of the overall refractive power of the cornea, based on a standardized conversion factor developed in the 1800s. Keratometry measures the radius of curvature of the anterior corneal surface, and the keratometric index of refraction takes into account the minus power of the posterior surface of the cornea as well as the actual index of refraction. This standardized factor is used so that a 7.5 mm radius of curvature corresponds to exactly 45 D (see Section 17).

Some texts say to fit 0.5 steeper than the low K. What does this mean?

It means that when fitting rigid contact lenses, one should choose a lens with a base curve 0.5 D greater than the lower K-reading. For example, if the K readings are 44.50/45.50, a contact lens with a curvature of 45.00 D (7.5 mm BC) should be chosen. If converting spectacle power to contact lens power, this "tear lens" must be subtracted from the calculated power when using a rigid lens. (Extra credit: When the corneal astigmatism exceeds 1.5 D, some experienced contact lens practitioners recommend fiiting steeper that the low K by one third of the total corneal astigmatism. Thus, with K readings of 44.00/47.00, there are 3 D of corneal astigmatism, and a lens with a base curve of 45 D would be chosen.)

After fitting a patient with rigid gas permeable lenses, you note that the fluorescein pattern indicates a low-riding lens with poor movement. What does this mean about the fit of the lens, and what are your options?

Steeply fit lenses tend to ride low and move poorly. They stick onto the cornea like a suction cup. You should consider decreasing the lens diameter, making the lens flatter, or (if this is an option) making the lens thinner.

One of your favorite patients is traveling on business and calls your office after her luggage has been stolen. Your staff can't find her chart, but the 36-year-old executive knows the type of contact lens she wears. She also knows both her spectacle and contact lens prescription, but can't remember which is which. One is a 5.50 and the other is 5.00, but she can't remember the sign. What would you ask her and what power contact lenses do you order?

Ask her whether she can read without her glasses. If she can, she is myopic; the contact lens is a -5.00 D, and the spectacle lens is -5.50 D. If she can't read without her glasses, she is likely hyperopic, and her spectacle prescription is +5.00 D. The contact lens prescription must be +5.50 D (see Chapter 6, *Lens Effectivity and Vertex Distance*).

The K-readings of a prospective contact lens wearer are 42.50/44.75 @ 85, but refraction indicates 20/15 visual acuity with -3.50 sphere spectacle correction. Since he has 2.25 D of corneal cylinder, should you prescribe a hard or soft lens? Should the lens be spherical or toric?

A spherical soft lens is probably the best choice. This patient's corneal cylinder is compensated for by lenticular cylinder. This is sometimes referred to as "favorable" or "complementary" astigmatism. A spherical hard lens is the worst choice, since it corrects the corneal cylinder but not the lenticular cylinder and leaves the patient with a large uncorrected astigmatic error. Try a spherical soft lens first, since it will not change the corneal cylinder, and it is less expensive than a toric lens. If the spherical soft lens somehow changes the balance between corneal and lenticular astigmatism, you may have to use a toric soft lens to correct the residual astigmatism.

An aphakic man wearing +12.00 D contact lenses requests glasses for backup. What power spectacle is required, if the vertex distance will be 10 mm? What about for a woman wearing -12.00 D contact lenses?

This is a classic vertex distance question. First, locate the far point of the eye. The focal point of the corrective lens should match the far point of the eye. In this case, it is 1 / 12 D = 0.083 m = 83 mm from the cornea. The man is hyperopic, so the far point is behind the cornea. The new lens should have a focal point that matches the far point of the eye. The location of the new lens is 10 mm in front of the cornea, or 93 mm from the far point. The power, then, will be 1 / 0.093 m = +10.75 D.

For the woman with myopia, the far point is 83 mm *in front* of the cornea, or (83 – 10) = 73 mm from the spectacle plane, so the power would be 1 / 0.073 = -13.7 D (Rx = -13.75).

You have been caring for Abe Phakia, an 18-month-old boy, since shortly after he was born. He has posterior lentiglobus (lenticonus), but the opacity did not fill the pupil and you had so far delayed surgery. You return from vacation to find that your partner has performed a successful lensectomy on Abe. Your partner now asks you to manage the patient's contact lens fitting and amblyopia therapy. You prescribe a 9.00/13.5/+18.00 extended wear soft contact lens. After a week, his mother, Sue D. Phakia, returns complaining that the contact lens is always falling out. What is the most likely problem, and what are your options?

The fit of the lens is probably too loose. To make the lens tighter, either decrease the base curve (try an 8.6) or increase the lens diameter. If you cannot obtain a soft lens that meets your parameters, consider a rigid gas permeable lens, even in children. Parents—and children—are surprisingly adept at learning how to remove and replace contact lenses on a daily basis.

11

Intraocular Lenses

Discuss the types of formulas in common use for calculation of intraocular lens power. Give an example of each type.

- *Theoretic* formulas are derived from optical principles and use assumed dimensions for certain parts of the eye (eg, Binkhorst, Colenbrander).
- *Empiric* formulas are derived by regression analysis from clinical results. A mathematical relationship between corneal curvature, axial length, and intraocular lens power can be derived (eg, SRK).

What does SRK stand for? What is it?

The SRK (Sanders, Retzlaff, and Kraft) formula is:

$$P = A - 2.5\,L - 0.9\,K$$

where P is the power for emmetropia, A is a lens constant, L is the axial length in millimeters, and K is the average corneal curvature in diopters. The lens constant is a unique constant related to lens type and manufacturer. It can be personalized by analysis of results with that lens.

Using the SRK formula, how much of an error in intraocular lens power would result from a 0.5 D mistake during keratometry? From a 0.5 mm mistake during axial length measurement?

- 0.5 D error in keratometry: 0.5 x 0.9 = 0.45 D, or about a half a diopter
- 0.5 mm error in axial length: 0.5 x 2.5 = 1.25 D!!

It should be obvious that small errors in axial length can translate into large errors in calculated lens power.

A patient presents with keratometry readings of 43 D and an axial length of 19 mm. Which IOL formula do you recommend?

Not the SRK or Binkhorst. Older theoretical formulas tend to overestimate the IOL power for short eyes, while SRK underestimates it. One of the newer-generation formulas should be used, as they are much more accurate with very long or very short eyes. Examples of newer formulas include the SRK II, SRK/T, and Holladay. These formulas are comparably accurate, though SRK II doesn't do as well in very long eyes. Only the SRK II is simple enough to calculate by hand. It is the same as the standard SRK formula, but the A constant is modified for long or short eyes, as in Table 11-1.

Hunter DG, West CE.
Last Minute Optics: A Concise Review of Optics, Refraction, and Contact Lenses, Second Edition (pp 65-70).
© 2010 SLACK Incorporated

Table 11-1

Value of the A Constant for the SRK II Formula

AXIAL LENGTH	MODIFIED A CONSTANT
Less than 20	Add 3
20 to 21	Add 2
21 to 22	Add 1
Greater than 24.5	Subtract 0.5

Note that in the previous example, if the SRK formula had recommended an IOL power of +32 D, the SRK II formula tells you to use +32 + 3 = +35 D.

SRK/T ("T" stands for "theoretical") and Holladay are extremely complicated theoretical formulas refined by empiric observations. A-scan units now contain computer software that calculates recommended IOL power based on one of the newer formulas. You should know which formula you use and be able to discuss whether you have had to customize it for your measured outcomes.

What is the disadvantage of a plano/convex IOL design (implanted with the convex surface closer to the cornea)?

A plano/convex IOL has more spherical aberration than a biconvex design. A biconvex IOL with a rear surface power that is stronger than the front surface power has very low spherical aberration.

Name an easy way to decrease the chance of making errors in axial length measurements.

There are many potential sources of error when measuring axial length and keratometry for IOL calculations. Measuring axial length in both eyes can decrease the chance of making a serious measurement error. If the 2 measurements are not close to equal, the measurements should be repeated.

A patient presents for preoperative evaluation for cataract surgery. The refraction in the cataractous eye is -9.50 sphere. She has been wearing a contact lens over that eye to correct the anisometropia. Keratometry readings in the eye are 43.00/42.50. What is a potential source of error in the corneal power calculations in this case?

The contact lens must be removed a minimum of 2 hours before keratometry. Errors caused by molding of the cornea can amount to 1 D. Some clinicians recommend removing the contact lens 2 *weeks* before keratometry, but others say 2 hours is enough time. Refractive surgeons will often ask that a patient go for a month or longer without contact lenses.

Should emmetropia be the goal in all cataract surgery?

Not usually. Induced anisometropia may produce intolerable aniseikonia, as well as inducing heterotropias. Take into consideration the status of the other eye (phakic vs. pseudophakic) and the preoperative refraction of the patient. For example, a moderate myope may wish to remain myopic so that she can continue to read without her glasses.

You are performing cataract surgery on Mr. Matt Ticulous, the husband of a successful malpractice attorney. After completing flawless phacoemulsification in record time, you load a +19.00 D lens into an injector and prepare to insert the lens in the capsular bag. To your horror, you realize that your new technician calculated the power for 1 D of *HYPERopia*, not the 1 D of *MYopia* that you requested. You remove the lens injector from the eye of your patient. Your IOL power calculator is 30 miles away at your satellite office, which is closed. What power lens do you insert in its place?

The general rule of thumb for normally sized eyes with an implant power of about +18.00 D is to change the calculated IOL power by 1.25 to 1.50 D for each diopter of desired ametropia. Therefore, if the calculated lens power for 1 D of hyperopia was +19.00, you should select a +21.50 or +22.00 D lens.

You complete the previous surgery after requesting and deftly inserting a +21.50 D intraocular lens into the capsular bag. While dictating the operative note, you realize that the circulating nurse handed you a +27.50 D lens. You quickly calculate that this mistake will leave the patient quite anisometropic. Will the patient be hyperopic or myopic in the operated eye? What do you tell the patient and his wife?

Myopic.

Consider the ethical considerations of mistakes and poor outcomes. What would you tell a patient after you made a mistake?

Although Mr. Ticulous had some difficulty adapting to his anisometropia, he returns to you for surgery in the other eye. You had intended to place your favorite lens implant, but surgery went very poorly and it was not possible to place any lens at all. Discuss the management of monocular aphakia, assuming the patient initially refuses more surgery.

If the aphakic eye has better vision, you may correct that eye with a contact lens or with aphakic spectacles and leave the other eye blurry. A secondary intraocular lens may be placed in the eye when the patient is ready.

If both eyes have good vision, a contact lens may be used for the aphakic eye. Had the patient been phakic in the other eye, and that eye developed a cataract, the patient could be left a bilateral aphake with subsequent contact lens or spectacle correction. This is rarely accepted by patients today; secondary intraocular lens insertion in both eyes will be appropriate.

Mr. Ticulous returns for his 23rd postoperative visit, 6 months following his first cataract extraction. He complains of decreased visual acuity in his pseudophakic eye. Best corrected visual acuity has dropped from 20/25 to 20/40. He also notes glare. How do you proceed?

First, recheck the refraction, and check pinhole acuity with best correction in place. Look at the IOL position before dilating the pupil. Are possible sources of glare visible (IOL defects, a centration hole, or an IOL edge)? Is there a posterior capsule opacity? Corneal decompensation? Look at the retina using your direct ophthalmoscope to evaluate the contribution of anterior segment opacities to visual acuity. Also, evaluate the retinoscopic reflex to identify irregularities or opacities in the media. Dilate the pupil and look again at the IOL and posterior capsule. If the front of the eye can't explain the decrease in visual acuity, direct your attention to the retina and beyond.

Five months later, during his 35th postoperative visit, Mr. Ticulous is ready for you to insert a secondary IOL in his aphakic eye. You select an IOL power that should leave the patient somewhat myopic and suture an IOL into the ciliary sulcus. Postoperatively, he returns with more myopia than expected, 4 D of astigmatism, and vertical diplopia. What may have happened?

The sutured IOL may have become dislocated. Decentration of the lens may be causing diplopia from prismatic effects, rotation of the lens may be causing astigmatism, and anterior dislocation of the lens may be causing excess myopia.

Dr. Allie Thumbs, a neurosurgeon, returns to your office for her first postoperative visit the day after uncomplicated cataract surgery with insertion of a 3-piece posterior chamber IOL. She feels well, but reports that she had a bad fall getting out of the tub this morning. Your tech reports counting fingers vision at 1 foot. On slit lamp exam, you note that the lens implant is not where you placed it. The capsular bag has ruptured from the trauma, and the lens implant is slightly posterior to its previous position. Assuming no rotation and no decentration of the lens, would you expect Dr. Thumbs to be more myopic or more hyperopic than she was just after surgery?

More hyperopic. The posteriorly displaced IOL pushes its focal point with it, decreasing its "effectivity." This renders the optics of the eye less powerful, and makes Dr. Thumbs more hyperopic than she was the day before.

The opposite would be true if Dr. Thumbs' IOL was anteriorly displaced and held in position by iris capture.

12

Refractive Surgery

Corey K. Tappia presents to your office seeking refractive surgery. On examination you find that the left pupil is slightly decentered. He says, "Oh, yeah, that's my devil eye—it's always been that way. Is it going to be a problem for me, doctor?" How do you center your refractive procedure in this case?

The key is to understand that the only region of the cornea that participates in the refraction of light is the portion that overlies the image of the pupil (the "entrance pupil"), even in extreme cases of corectopia. If you were to center the refractive procedure on the center of the cornea in this case, the edge of your ablation might overlap with the entrance pupil and cause an unpredictable result with potentially severe glare. After evaluating the patient for possible medical causes of corectopia, you should be able to proceed with the refractive procedure, centering on the center of the undilated pupil.

How does refractive surgery on the cornea correct myopia? What is the basic principle conceptually, and how is it expressed mathematically?

In myopia, either the eye is too long or the corneal refractive power is too strong. In refractive surgery, the cornea is flattened in the region where traversing light will enter the pupil. This decreases its refractive power by increasing the radius of curvature in the equation, $D = (n' - n) / r$, such that if "r" is larger (and the cornea flatter), D will be smaller. Flattening can be achieved by radial incisions that steepen the periphery (with compensatory flattening in the center), or by sculpting the corneal surface with an excimer laser.

How does hyperopic refractive surgery work?

In hyperopia, the cornea is not strong enough for a given axial length, so the procedure must somehow increase the corneal curvature over the pupil. This can be accomplished with excimer laser sculpting as noted previously, or by using conductive keratoplasty to "cook" the corneal periphery, thus creating a band of flat, shrunken cornea peripherally that causes compensatory steepening centrally.

How can a laser be used to correct astigmastism? If keratometry suggests that the 45° meridian is steepest, what sort of ablation would be performed?

The key for correcting astigmatism is to alter the corneal stroma in an elliptical pattern. One way to achieve this is to flatten the cornea, preferentially in the steepest meridian. For this patient, an elliptical ablation would need to take place in the 45° meridian. Alternatively, a plus cylinder technique could be used to steepen the cornea in the 135° meridian.

Hunter DG, West CE.
Last Minute Optics: A Concise Review of Optics, Refraction, and Contact Lenses, Second Edition (pp 71-74).
© 2010 SLACK Incorporated

Mr. Lee Tigious presents to you seeking laser vision correction, but he has been told that he has "the Stigmatism" and he worries that it is incurable. Refraction is:

OD: -3.75 +1.25 x 90 => 20/15

OS: -3.75 +1.25 x 90 => 20/15

Keratometry measures:

OD: 45.00 @ 90 / 46.00 @ 180

OS: 45.00 @ 90 / 46.00 @ 180

What kind of astigmatism does he have? Is the astigmatism going to be a problem in this case?

In this case, the power cross places -3.75 on the 90° meridian and -2.50 on the 180° meridian. Both numbers are negative, so this is compound myopic astigmatism. The axis of plus (correcting) cylinder is at 90°, so it is with-the-rule.

The keratometry results indicate that there is 1 D of corneal astigmatism, so this seems at first glance to correspond roughly with the corneal astigmatism. But does it? The cornea is steeper in the 180° meridian. This would be corrected by providing about a diopter of plus power in the 90° meridian—plus cylinder axis 180! That is, given these keratometry measurements, you should expect the corrective lenses to be "<some sphere> +1.00 x 180." You have uncovered a rare case of lenticular astigmatism that is neutralized (and then overcompensated for) by complementary corneal astigmatism. If your refractive surgery eliminates the corneal astigmatism, the patient will end up with over 2 diopters of uncorrected lenticular astigmatism.

Since refractive surgery is generally based on the subjective refraction, the lenticular astigmatism should not be a problem—at least not for now. However, should the patient develop cataracts, removing the complementary lenticular astigmatism will result in unexpected, significant postoperative astigmatism and an unhappy patient.

What sort of optical aberrations might be brought out by a laser ablation zone that is too small?

If the edge of the ablation zone is within the entrance pupil of the eye, light rays traversing this "aperture" are subject to diffraction that can cause glare. In addition, there are 2 different refractive powers within the pupil, which may lead to multifocal optics and monocular diplopia (or even triplopia in some cases.)

Name some optical aberrations that can be induced by refractive surgery.

In addition to diffraction and spherical aberration noted previously, there can be increased scattering (from an irregular ablation or stromal haze), coma (from a decentered procedure), and spherical aberration (from an alteration in the overall curvature of the cornea). These various aberrations can lead to starbursts, monocular diplopia, flare (like a comet), and halos.

13

Magnification and Telescopes

Calculate the transverse magnification for the images in Figure 2-7 (page 9).

A. X = 100 cm

B. X = 50 cm

C. X = 250 mm

D. X = 12.5 cm

E. X = 0.011 m

Is each image inverted or upright?

Transverse magnification = (image size) / (object size) = (image distance) / (object distance). For Part A, magnification = (14 / 100) = 0.14 (minification). To determine whether it is inverted, draw the central ray and locate the object and image with respect to the lens (Figure 13-1).

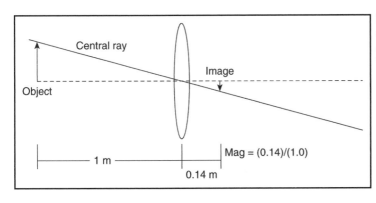

Figure 13-1. Transverse magnification of image in Figure 2-7, part A.

A: 0.14, inverted (0.14 / 1)
B: 0.34, inverted (0.17 / 0.50)
C: 1.00, inverted (0.25 / 0.25)
D: Undefined, image is at infinity
E: 1.09, upright (0.012 / 0.011)

Hunter DG, West CE.
Last Minute Optics: A Concise Review of Optics, Refraction, and Contact Lenses, Second Edition (pp 75-80).
© 2010 SLACK Incorporated

Calculate the transverse magnification for the images in Figure 2-7, now changing the power of the lens to -8 D. Is each image inverted or upright? Use the same approach as in the previous question.

A: 0.11, upright (0.11 / 1)
B: 0.20, upright (0.10 / 0.50)
C: 0.33, upright (0.083 / 0.25)
D: 0.50, upright (0.0625 / 0.125)
E: 0.91, upright (0.010 / 0.011)

Calculate the transverse magnification for the images in Figure 2-8 (page 10). Is each image inverted or upright?

For multiple lens systems, work sequentially from left to right to calculate transverse magnification and to draw the central rays. Remember, the image of the first lens is the object of the second (Figure 13-2).

Figure 13-2. Step 1: Magnification of image in Figure 2-8A.

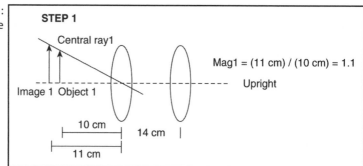

From the first lens system we have a magnification of 1.1. Image 1 will now become the object of the second lens system (Object 2). It is (11 + 14) = 25 cm away from the second lens (Figure 13-3).

Figure 13-3. Step 2: Magnification of image in Figure 2-8A.

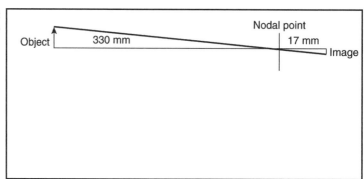

From the second lens, the transverse magnification is 4.0. The magnification for the entire system is (1.1) x (4.0) = 4.4. From the ray tracing, the image is inverted.

From Figure 2-8, Part A

       Step 1:   Magnification 1.1, upright
       Step 2:   Magnification 4.0, inverted
       Final: Magnification 4.4, inverted

From Figure 2-8, Part B
>   Step 1:   Magnification 2.0, upright (20 / 10 = 2.0)
>   Step 2:   Magnification 1.6, upright (53 / 34 = 1.6)
>   Final: Magnification 3.2, upright (2.0 x 1.6)

From Figure 2-8, Part C
>   Step 1:   Magnification 0.9, upright (9 / 10 = 0.9)
>   Step 2:   Magnification 6.1, inverted (1.4 / 0.23)
>   Final: Magnification 5.5, inverted (6.1 x 0.9)

What is the magnification of a 20 D condensing lens when used as a simple magnifier?

5X.

A simple magnifier is a plus lens used to increase the angular subtense of a near object. The magnifying power is always expressed relative to an arbitrary reference distance. This was established long ago as 25 cm. The angular magnification is equal to the dioptric power of the lens divided by 4. Thus, a 20 D condensing lens has a magnification of (20 / 4) = 5X when used as a simple magnifier.

What is the magnification of a 20 D condensing lens when used as a simple magnifier if the reference distance is changed to 40 cm?

8X.

For a reference distance of 25 cm, the magnification is D / 4 = D x (0.25 m). For a reference distance of 40 cm, the magnification is D x 0.4 m = D / 2.5 = 8X. The same lens is considered to have a different magnification depending on what viewing distance that magnification is compared with. Note that for a reference distance of 1 km, the magnification is D x 1000 m = 20,000X. That is, an object viewed through a 20 D lens used as a simple magnifier takes up 20,000 times more of your visual field than it would if it were held 1 km away and viewed without magnification.

You and a stowaway are stranded on a lost planet with your trial lens set, but only a few lenses survived the shipwreck. You are left with a -20.00 sphere, +4.00 sphere, +5.00 sphere, and a +20.00 sphere. You build a viewing device to search the horizon for ships using the -20.00 and the +4.00 lens. The stowaway, Dr. Smith, uses the +20.00 and the +5.00 lenses. Dr. Smith complains that his telescope is inferior.

A.   What did each of you build, how did you position the lenses, and why is Dr. Smith plotting to steal your telescope?

You used the -20 D lens as the eyepiece and the +4 D lens as the objective of a Galilean telescope. The secondary focal point of the plus lens (1 / 4 = 0.25 m) should coincide with the primary focal point of the minus lens (1 / 20 = 0.05 m), thus the lenses are 25 cm – 5 cm = 20 cm apart.

Dr. Smith built the second telescope (astronomical) using the +20 as the eyepiece and the +5 D lens as the objective. The secondary focal point of the objective needed to coincide with the primary focal point of the eyepiece lens, so he positioned them 5 cm + 20 cm = 25 cm apart. Dr. Smith does not like having to stretch his arms the additional 5 cm.

B.   Which telescope above will provide you with more magnification?

The angular magnification of a telescope is equal to the power of the eyepiece divided by the power of the objective. The magnification of the Galilean telescope is 20 / 4 = 5X, and the magnification of the astronomical telescope is 20 / 5 = 4X. Thus, Dr. Smith's telescope will provide less magnification.

C. Will the telescopes have upright or inverted images?

The Galilean telescope will provide an upright image of approaching ships, while Dr. Smith's astronomical telescope will produce an inverted image. For all of these reasons, Dr. Smith is plotting to steal your superior telescope.

How is the Galilean telescope modified when used as a surgical loupe?

The binocular surgical loupe is just a short Galilean telescope fitted with an add to bring the working distance in from infinity. Powerful lenses are used so the telescopes won't be too long; a +25 D objective and a -50 D eyepiece provides 2X magnification. To calculate the additional add, take the reciprocal of the working distance in meters.

How long is the 2X Galilean telescope described previously? What if it were made using a +5 D objective and a -10 D eyepiece?

The focal length of the -50 D lens is 1 / 50 = 2 cm. The +25 D lens has a 1 / 25 = 4 cm focal length. Thus, the telescope is 4 – 2 = 2 cm long. The +5/-10 telescope is 20 – 10 = 10 cm long.

What is an "error lens" and why is it a useful concept?

When an eye is myopic, it can be thought of as "too powerful" because light rays are brought into focus in front of the retina. One way to simulate an eye that is "too powerful" is to insert extra plus power in the eye. That is, if Dr. West were emmetropic, and Dr. Hunter surgically inserted a +10 D iris claw anterior chamber IOL in her eye without disturbing her natural lens, she would end up as a myope requiring slightly less than -10 D of contact lens correction to see clearly at distance. The +10 D surgically implanted IOL can be considered an "error lens."

Similarly, the error lens in a hyperope has minus power. The error lens concept is especially useful when calculating magnification caused by correction of ametropia.

Why do patients respond that things look "smaller" when they are "overminused" during the course of a refraction?

Too much minus in the trial frame or phoropter effectively forms the eyepiece of a Galilean telescope, and the extra plus power in the eye (the "error lens") forms the objective. The patient's retina is thus looking backwards through a Galilean telescope, and the letters are minified.

What is aniseikonia and what causes it?

Aniseikonia is a difference in perceived image size between the two eyes. Most normal adults can tolerate a 3% to 8% difference in image size, and children can tolerate even more. Aniseikonia is most commonly caused by unequal refractive errors (monocular aphakia or pseudophakic surprises, for example), but is also found with retinal problems and occipital lobe lesions. A handy rule-of-thumb for spectacle correction is that each diopter changes the retinal image size by about 2% (plus lenses magnify, minus lenses minify).

What is iseikonia?

Iseikonia is when the perceived images are the same size.

What is anisometropia?

Anisemetropia is a difference in refractive error between the two eyes. The popular definition of clinically significant anisometropia (2 D of spherical difference between the eyes) is arbitrary. The dioptric amount of anisometropia is not the determining factor in how much anisometropia a patient finds tolerable. Other factors are the type of anisometropia, its duration, the patient's age, binocularity and fusional potential, and the type of correction (spectacles versus contact lenses) a patient wears.

Does Knapp's Rule apply to axial or refractive myopia or both?

Knapp's rule states that the proper corrective lens placed at the anterior focal point of an eye will produce retinal images of the same size no matter what amount of axial ametropia exists. Two problems prevent this from being strictly applied in clinical practice: 1) ametropia is almost never purely axial, and 2) it is impractical to set a vertex distance of 15 to 16 mm for spectacle correction (see the discussion of model eyes). Furthermore, the retina in the myopic eye of a unilateral high myope is stretched; this increase in separation of photoreceptors can cause confounding changes in effective magnification.

Ansel Metropia is 65-year-old woman who has been your patient for the past 10 years. She has a history of amblyopia successfully treated with patching in childhood. Recently she presented with clinically significant nuclear sclerosis in both eyes, measuring:

OD: -9.00 +0.50 x 116 => 20/50

OS: -6.50 +0.75 x 70 => 20/100

You performed your usual flawless cataract surgery in each eye and, thanks to your exquisite biometry and IOL calculations, the 4-week postoperative refraction is:

OD: -1.00 sphere => 20/20

OS: -1.00 sphere => 20/15

You glance at the numbers on the chart and enter the exam room expecting cards, flowers, and hugs. Instead, Miss Metropia is scowling and her boyfriend is wringing his hands.

"Everything is double," she growls.

"She c-cant do anything," whimpers her boyfriend, "She just sits on the couch at home. I even had to drive today."

You perform a quick alternating cover test and find no hint of strabismus. What went wrong? What can you do now?

The problem is that Ansel is suffering from an unusual case of aniseikonia. Recalling that there was some anisometropia before surgery, you review her record and see that 10 years ago, she measured:

OD: -8.00 +0.50 x 116 => 20/20

OS: -2.50 +0.75 x 70 => 20/15

You also not that she had excellent stereopsis before surgery. Thus, she had 5.5 D of anisometropia that was so well treated in childhood that her brain had fully adapted to the difference in image sizes. As the cataract progressed in the left eye, there was asymmetric myopic shift, causing the anisometropia to appear to be less severe. By eliminating the anisometropia with surgery, you enlarged the image on the left side by (2% x 5.5 D) = 11%. Most adults can tolerate 3 to 4 D of anisometropia (6% to 8% aniseikonia), but few can tolerate 11%.

To treat, you could place a +4 D contact lens on the right eye, equivalent to an error lens that would make the right eye require a -5 D spectacle lens versus the -1 D spectacle on the left, thus restoring the difference between eyes. Trial and error could help determine the lowest plus power contact lens necessary to eliminate the aniseikonia. Once this was known, if the patient did not wish to be dependent on contact lenses, you could offer IOL exchange or refractive surgery.

14

Low Vision

When should low vision aids be given?

When a patient's visual needs exceed his or her visual capabilities. This may occur with a visual acuity of 20/40 in one patient, and 20/200 in another.

Describe the steps you would take in the evaluation of a 52-year-old man with low vision.

- *History:* Duration and course of visual loss may influence his motivation and acceptance of aids. What were his habits prior to visual deterioration? What are his current occupational and avocational needs? Does he have realistic expectations of what you can do for him? Does he have any other physical limitations (ie, tremor, deafness)? Is he technophobic? Does he have any low vision aids that are currently useful or used to be of use?

- *Examination:* This should include a precise refraction, with distance and near acuity. Visual acuity should be measured with a low vision chart, or by moving the patient closer to the chart. Does the patient have any accommodation? Other findings that may influence the type of aid prescribed should be noted, ie, nystagmus, photophobia, aniridia, or visual field defects.

- *Trial of a variety of low vision aids:* Which aids seems to be most appropriate for his needs? Most patients will need more than one low vision aid.

A patient presents to your office unable to read the big "E" on the eye chart. How do you quantify the patient's visual acuity in this case?

If your office does not carry low vision cards designed specifically to assess visual acuity in low vision patients, you can simply move the patient closer to the eye chart. The visual acuity changes proportionally. That is, if the patient can read the 20/100 line from 5 feet away, then visual acuity is 5/100 (which converts to 20/400).

Note that if a patient has nystagmus with a latent component, monocular occlusion will reduce visual acuity. In such cases, fog the other eye with a plus lens rather than covering with an occluder, and measure visual acuity under binocular conditions as well. Nystagmus patients with a null point may also require a compensatory head posture for best visual acuity.

You should also measure near visual acuity (under both monocular and binocular conditions) in low vision patients.

Hunter DG, West CE.
Last Minute Optics: A Concise Review of Optics, Refraction, and Contact Lenses, Second Edition (pp 81-84).
© 2010 SLACK Incorporated

Discuss Kestenbaum's Rule.

Kestenbaum's Rule can help you estimate the strength of a near add for a low vision patient. At a reading distance of 40 cm (16 in), newspaper print (J5, 8 point, 1M) requires visual acuity of approximately 20/50. This reading distance (0.4 meters) just *happens* to require an add of +2.50 D (1 / 0.4), which is the reciprocal of the Snellen acuity. A patient with 20/120 distance acuity would be estimated to require a near add of (120 / 20) = +6.00. This is just a starting point—you still need to put the lenses in trial frames and hand the patient a newspaper.

An 11-year-old honor student is discovered to have an hereditary optic atrophy. Her best-corrected visual acuity is 20/80 OU with -3.25 sphere OU. What sort of add would you give her?

Not necessarily any. As long as she has normal accommodation, an 11-year-old child should be able to accommodate well enough to hold objects sufficiently close to achieve high levels of magnification on her own. Her accommodative abilities can be assessed by allowing her to hold the near card as closely as she likes while you check near visual acuity and perform dynamic retinoscopy.

Is there a difference between a stand and a hand magnifier?

Both are aids for near, but each has its advantages and disadvantages. Hand magnifiers are inexpensive, easily obtained, easy to carry, and have a variable eye-to-hand distance. However, they have a small field of view when held farther from the eye and may be difficult for patients with strokes, tremor, or arthritis. Stand magnifiers are easier for patients with strokes, tremor, or arthritis, but are bulkier and harder to carry in a pocket or purse. Stand magnifiers come with fixed distance or with adjustable height. The fixed stand magnifiers are designed to be used with a reading add.

A 57-year-old dentist has early onset macular degeneration and a best-corrected visual acuity of 20/50. He has 3 children in college and alimony payments to make. Because he allowed his disability policy to lapse, he needs to keep his practice going to make ends meet. Should you prescribe a hand magnifier, or are there other options?

A hand magnifier is clearly inappropriate, as the dentist needs both hands free for his work. Higher power adds with base-in prism might be considered, but only momentarily, as he needs a longer working distance when working in his patients' mouths. (The base-in prism is helpful with high power adds because the short working distance requires excessive convergence.) A surgical loupe would be the best bet, as he could get higher magnification with a reasonable working distance.

When is a telescope useful for a low vision patient?

A telescope is used when a patient needs to have better visual acuity at distance than spectacle correction allows. Telescopes can be spectacle-mounted (binocular or monocular) or hand-held. Patients typically use them to see bus signs or street signs. In some cases and in some states, telescopes will allow patients to qualify for a driver's license if their visual field is full—this is best achieved with bioptic style, spectacle-mounted telescope used only to read street signs so the peripheral vision is otherwise-preserved. Spectacle-mounted telescopes are cosmetically obvious and heavy, and all telescopes have a limited field of view. Distance telescopes can be used as a near aid with the addition of a cap on plus lens.

Discuss the advantages and limitations of closed circuit television (CCTV) and related devices.

CCTV devices provide high magnification and high or reverse contrast, but they are expensive and not very portable. The reading material must be moved around in the field of the camera, and this can be difficult for people with problems with manual dexterity and when the magnification is very high.

Other advanced technologies are emerging for low vision patients. For example, Ray Kurzweil has developed a cell phone application that allows a patient to photograph a sign with the phone camera, process the image for text, and convert the text to speech.

A computerized scanner requires much less dexterity on the part of the user than standard CCTV. Entire pages (even with multiple columns) are scanned into a computer, processed, and displayed as a single line of text that streams across the screen at a user-controlled rate.

Discuss nonoptical modalities in the treatment of low vision.

Large print books and subscriptions, templates, magic markers, signature guides, computer programs, high contrast materials, good lighting, large-print playing cards, as well as large print telephones, watches, and timers are non-optical modalities. Absorptive lenses can help those people who are bothered by glare (eg, albinos, patients with aniridia) or who see better in lower light levels (eg, congenital achromatopsia patients do best with a very dark *red* lens). Yellow-tinted lenses remove blue light; since even normal subjects have very few blue cones, the lenses can decrease glare and scattering caused by blue light without impairing central visual acuity.

15

Mirrors

A frugal, 160-cm-tall ophthalmologist wishes to purchase a mirror from the new superstore, "Mirrors by the Meter." She will mount the mirror on the wall and use it to view her whole self each morning before heading to the office. She must choose among the 1, 2, and 3 m mirrors. Which is the shortest (thus, the least expensive) mirror she can purchase to get the job done? Is her image real or virtual?

She should purchase the 1 m mirror. The light ray heading from the base of her heel up to her eye will reflect off of the mirror at a point 80 cm above shoe level. No mirror is required below that point. If the top of the mirror is precisely at eye level, only an 80 cm mirror is needed, plus a few extra cm to allow her to view her bouffant hairdo. (If she has very pointy shoes that extend forward, she will require a few more cm [beyond 80] at the bottom of the mirror to see the tips of her shoes.)

A plano mirror just reverses the direction of the light without changing vergence, and therefore it has zero power. The image formed of a real object by a plano mirror is always virtual, erect, and of the same size as the object.

What part of the eye is used as a convex mirror in ophthalmic instrumentation?

The cornea (actually the tear film) is used as a convex mirror by keratometers.

Ginger, a model and actress, is checking her eye makeup and notices a speck between 2 lashes. She is 10 cm from a plano mirror. How far away is the image of the speck from the mirror? From Ginger's eye?

The vergence entering the mirror is $U = 1 / (0.1 \text{ m}) = -10$ D. Vergence added by the plano mirror (D) is zero. Vergence leaving the mirror is $(0 + -10) = -10$ D. Therefore the image is $1 / (10 \text{ D}) = 0.1 \text{ m} = 10 \text{ cm}$ from the mirror, or 20 cm from Ginger's eye.

Concerned about the speck, Ginger flips her mirror over and looks into a concave mirror on the other side of the plano mirror. The concave mirror has a radius of curvature of 50 cm. How far away from Ginger's eye is the magnified image of the speck, which she now realizes is a louse? Is it inverted or upright? What is the transverse and axial magnification of the louse?

Ginger has not moved (yet), so the vergence entering the mirror is still $U = 1 / (0.1 \text{ m}) = -10$ D. Vergence added by the concave mirror (D) is $2 / r = 2 / (0.5 \text{ m}) = +4$ D. Vergence

Hunter DG, West CE.
Last Minute Optics: A Concise Review of Optics, Refraction, and Contact Lenses, Second Edition (pp 85-88).
© 2010 SLACK Incorporated

leaving the mirror is therefore (-10 D + 4 D) = (-6 D) (Figure 15-1). Therefore the image is 1 / (6 D) = 0.167 m = 16.7 cm from the mirror. The light is still diverging after it is reflected by the mirror, so the image is a virtual image located by finding the intersection of imaginary extensions of light rays 16.7 cm to the right of the mirror, or 26.7 cm from Ginger's eye. Draw a central ray through the radius of curvature of the mirror to find that the image is upright and magnified. The transverse magnification is (image distance) / (object distance) = 16.7 / 10 = 1.67x (Figure 15-2).

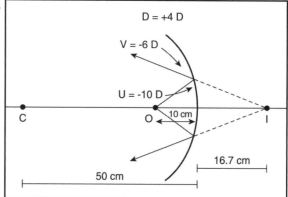

Figure 15-1. Locate the image of the speck using U + D = V.

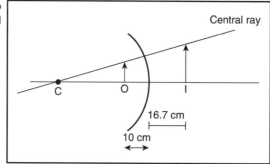

Figure 15-2. Draw the central ray to determine that the image is upright and magnified.

(Note: To calculate magnification, you may use image and object distance from the mirror or image and object distance from the center of curvature—it comes out the same either way.) Axial magnification is (transverse magnification)^2 = 1.67^2 = 2.8x. So the image appears to be 1.67 times taller and 2.8 times deeper than the object.

16

Prisms and Diplopia

Calculate the prismatic displacement at positions A through E. Is the light ray displaced up or down (Figure 16-1)?

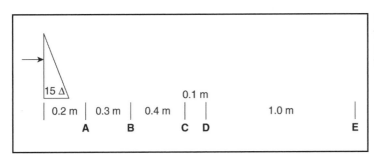

Figure 16-1. Calculate the prismatic displacement at positions A through E.

The light ray will be displaced down, toward the base. It will be displaced 15 cm at 1 m by the 15 prism diopters (PD, by definition of a prism diopter). Therefore at 0.2 m, for example, it will be displaced (Figure 16-2):

$$\frac{0.2\ m}{1\ m} * 15cm = 3cm$$

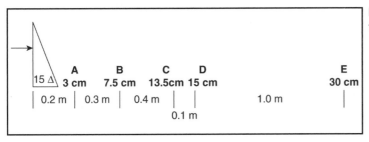

Figure 16-2. Solution to Figure 16-1.

Hunter DG, West CE.
Last Minute Optics: A Concise Review of Optics, Refraction, and Contact Lenses, Second Edition (pp 89-98).
© 2010 SLACK Incorporated

A student attaches a base down, 10 PD prism to the front of a slide projector lens. The lens is 6 m from the screen.

A. Which way does the image of the projected slide move?

The image moves down (Figure 16-3).

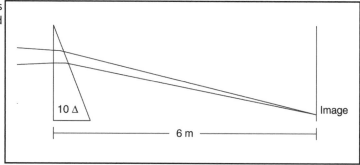

Figure 16-3. Prisms bend light rays toward the base.

B. How far does it move?

60 cm.

$$\left(6 \text{ m} * \frac{10 \text{ cm}}{1 \text{ m}} = 60 \text{ cm} \right)$$

C. Is the image real or virtual?

This is a real image. It is on the same side of the prism as the outgoing rays and is not located by imaginary extensions of light rays.

What are the indications for prisms in spectacles?

Prism is used to restore binocularity in a patient with constant strabismus and to ease asthenopia in a patient with a phoria or intermittent tropia. Consider prescribing prisms for patients with small, comitant deviations. Larger deviations are harder to treat with prisms, simply because of the weight of and chromatic aberration produced by spectacle prisms. Consider a trial of Fresnel prisms before grinding the prism into the spectacle lens, but be aware that some patients cannot tolerate the glare and decreased visual acuity caused by the Fresnel prism.

For example, patients with intermittent exotropia, convergence insufficiency (XT at near), divergence insufficiency (ET at distance), anisometropia in the reading position (with vertical phoria), or acquired strabismus after intraocular surgery may all benefit from prism therapy. Base-out prism might also be given to a child with acquired esotropia prior to surgery, a form of prism adaptation that is used to bring out the "true" amount of esotropia.

How should prisms be written in prescriptions?

Always specify the amount and direction of the prism, and the lens or lenses into which the prism should be incorporated. For comitant strabismus, it is best to split the prism, otherwise one lens will be very heavy compared with the other. You can specify the amount of prism in each eye, or you can specify the total amount and then add "okay to split the prism" on the prescription. For incomitant strabismus, it is best to try to place the prism over the eye with limited movements in order to bring the image to the eye, rather than requiring the oculomotor system to try to bring the eye to the image, which may induce a secondary deviation.

For combinations of horizontal and vertical prism, opticians are perfectly happy for you to write out the horizontal and vertical prism separately; they will then use that information to grind in the prism, with the net result being an oblique prism. However if the goal is to apply a Fresnel press-on prism over one eye, then it is important to be exact about the direction. For example, if you write "6 PD @ 25°, right eye," the optician could place the prism base-up and out in the 25° meridian OR base-down and in on the 25° meridian (Figure 16-4). Therefore you should be exact (eg, "6 PD base up-and-out in the 25° meridian").

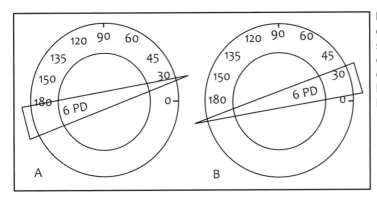

Figure 16-4. Ambiguous result of prescribing "6 PD @ 25°" can lead to A) base down-and-out, or B) base up-and-in, both in the 25° meridian.

What is Prentice's Rule?

PD = h D

where D is the dioptric power of the lens, h is the distance from the optical center of the lens in centimeters, and PD is the number of prism diopters. It determines the induced prismatic power of a lens when light passes through any point away from its optical center.

A 7 PD prism is placed base-up and out in the 45° meridian over the left eye of an orthotropic patient. The patient complains of diplopia.

A. What is the induced vertical prismatic power?

5 PD base-up, left eye.

The first step is to break the problem down into its vertical and horizontal components. Since the angle is 45°, the 2 components are equal (Figure 16-5).

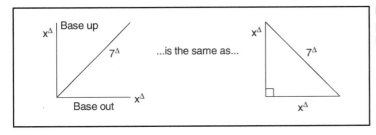

Figure 16-5. Prism power in oblique meridians.

This calls for the Pythagorean theorem:

$$x^2 + x^2 = 7^2 = 49 \text{ or } 2x^2 = 49. \text{ Thus, } x = \sqrt{25} = 5$$

The vertical component of induced prism power is thus 5 PD, oriented base-up over the left eye.

B. What is the horizontal component of the prism power?

5 PD base out. Use the same reasoning as with previous answer.

C. What types of strabismus appear to be present when an alternate cover test is performed?

A left hypertropia and an exotropia. To understand the induced vertical deviation, imagine that the left and right eyes are looking at an object (with the right eye fixing on the object). Trace a ray of light from the fovea of the left eye through the base-up prism and it will be deflected above the object. In other words, the left eye is looking higher, producing a left hyperdeviation. Another way to think of it is that if a base-up prism is placed over the left eye, it would be corrected by a base-down prism, which we know from clinical experience is used to treat a hyperdeviation in that eye. Similar reasoning explains the exotropia.

D. What prism could you place over the right eye to eliminate the diplopia?

7 PD, base-up and in. To correct a left hypertropia (right hypotropia), a base-up prism is used over the right eye. To correct an exotropia, a base-in prism is used over either eye.

How does image jump differ from image displacement?

Image jump is seen because of the *sudden* introduction of prismatic power at the top of the bifocal segment. With round top segments, the sudden introduction of base down prism causes an object to "jump" upward as the eye moves down. If the optical center of the bifocal is at the top of the segment, as in the Franklin (Executive) type, there is no image jump. (No image jump is seen with progressive addition lenses.)

Image displacement is produced by the total prismatic power acting in the reading position, and can be minimized when the prismatic effect of the bifocal is opposite to that of the distance segment.

Myopic patients should be given flat top segments (or Franklin [Executive] style) to minimize both image jump and image displacement. For *hyperopic patients*, the choice is not as easy, for it depends on what might be most bothersome to that particular patient. Image jump might theoretically be more bothersome to someone who switches distance-near fixation frequently (waiters, receptionists, shopkeepers); however, the switch from distance to near fixation is often so rapid that image jump goes unnoticed. Image displacement might bother a desk worker (lawyer, accountant, graphic artist) more. For hyperopes, round tops minimize image displacement and maximize image jump; flat top or Franklin-type segments minimize image jump while maximizing image displacement.

In practice, many opticians give flat top segments to myopes and hyperopes because flat top segments are easier and cheaper to assemble. As long as a patient is not complaining, this is acceptable; just remember to consider these issues when you encounter a patient who does not like his or her new bifocals.

After you perform uneventful bilateral cataract surgery on a local malpractice lawyer, Ruth Leslie Lawless, her refraction is OD - 4.00 sphere = 20/15 and OS - 1.00 sphere = 20/15; with a +2.50 add she can read J1+ at a comfortable distance with each eye. She is orthophoric at distance and does not seem to be bothered by aniseikonia, but complains bitterly of vertical diplopia in downgaze with her new glasses.

A. Ms. Lawless normally reads 0.7 cm below the optical centers of her Executive style bifocals. How much prism is induced in the reading position?

Using Prentice's rule, in the right eye there is (-1.50 D) x (0.7 cm) = 1.05 PD of prism (base-down, since the patient is myopic). In the left eye there is (+1.50 D) x (0.7 cm) = 1.05 PD, base-up. Base-down prism over one eye *adds* to base-up prism over the other eye when considering an induced phoria. The net prismatic effect is therefore (1.05) - (-1.05) = 2.1 PD base-down over the right eye (or 2.1 PD base-up over the left eye). The patient might be able to learn to fuse this amount of vertical phoria, but right now she needs to have something done.

B. Name some ways to limit the induced prismatic effects of her anisometropic correction in downgaze.

Many patients physiologically adapt or learn to fuse small vertical deviations, but if they can't, the following methods may help:

- Slab off prism (bicentric grinding)
- Contact lenses instead of glasses
- Lower both optical centers to compromise the vertical imbalance between distance and near
- Separate single vision glasses for distance and near
- Dissimilar segments
- Fresnel prism over bifocal segment

A 20-year-old jewelry designer who has never worn glasses before is given the following distance correction:

OD: -8.00 sphere

OS: -8.00 sphere

He does not require bifocals. The patient reads 12.5 mm below the optical center of each lens.

A. What is the induced prism in each eye?

10 PD base-down OU. For prismatic calculations, use the visual mnemonic shown in Figure 16-6 to remind you that minus lenses act like apex-to-apex prisms. (Similarly, plus lenses act like base-to-base prisms.) Use Prentice's rule: 1.25 cm x 8.00 D = 10 PD base-down OU.

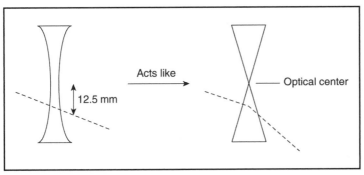

Figure 16-6. Visual mnemonic describing prismatic behavior of a minus lens.

B. If the reading position were 10 cm below the optical centers, how far would the image of an emerald held 0.5 m away appear to be displaced for each eye?

5 cm for each eye.

$$\left(10 \text{ cm } * \frac{0.5 \text{ m}}{1.0 \text{ m}} = 5 \text{ cm} \right)$$

C. If the patient tried to reach for the emerald with forceps, would he reach above or below it?

Above. The image is displaced upward, higher than the original object (Figure 16-7).

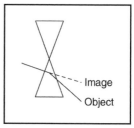

Figure 16-7. Image displacement in down gaze through a minus lens.

D. Is the image seen by the patient real or virtual?

Virtual. The image is located using imaginary extensions of the rays of light.

E. What vertical strabismic deviation would be measured in the reading position?

None! With base-down prism placed before both eyes, each eye sees the same amount of image displacement; thus, there is no discrepancy between the eyes.

F. What horizontal strabismic deviation would be measured if the patient looked 10 mm to the right of the optical centers?

None. Consider the diagram in Figure 16-8.

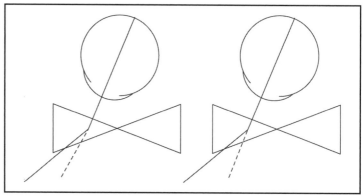

Figure 16-8. Image displacement in gaze right. The image is displaced equally in both eyes.

The image is displaced equally for both eyes. (Base-out prism over right eye, base-in prism over left eye). The image displacement is (10 cm) x (8 D) = 8 PD to the left for both eyes.

G. In the reading position, each eye also moves 3 mm nasally. What is the net horizontal prismatic effect of this eye position?

4.8 PD, base-in. In this case, the spectacles contribute base-in prism power of (0.3 cm) x (8 D) = (2.4 PD) base-in to each eye. The net prismatic effect is 4.8 PD of base-in prism (Figure 16-9).

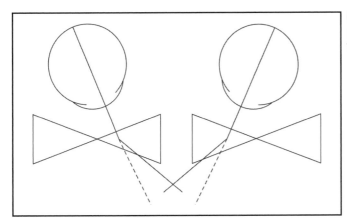

Figure 16-9. Image displacement in reading position.

Princess Rule of the Kingdom of Strabismonia is wearing +5.00 sphere OD and -5.00 sphere OS. She presents complaining of many years of diplopia. You measure a 4 PD esotropia. There is no problem with the spectacles, and you wish to prescribe prisms. Unfortunately, prisms are forbidden in the kingdom, a crime punishable by death (the King had been subjected to excessive orthoptic therapy during childhood). *Without prescribing a prism,* how could you move the optical centers of the spectacle lenses to treat the deviation in the primary position?

The desired effect is 4 PD of base-out prism. There is more than one correct way to achieve this. One approach is to displace each optical center 4 mm to the patient's right to give 2 PD of prismatic effect in each eye. Consider the diagram in Figure 16-10.

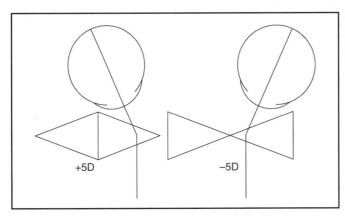

+5D   −5D

Figure 16-10. Intentional displacement of the optical centers to achieve prismatic power.

In the right eye, the (plus) lens should be decentered to the patient's right, for an induced prismatic deviation of (x cm) x (5 D) = 2 PD. Thus, x = 2 / 5 = 0.4 cm or 4 mm to the right. For the left eye, the (minus) lens should also be decentered to the right by 4 mm for a 2 PD base-out effect. The combination gives a total of 4 PD base-out. The problem in this case is that the diplopia will return when the patient looks to the right and is again using the optical centers of the spectacles.

Note that had the left lens been a plus lens, you would have had to move the optical center to the left to eliminate the diplopia in primary position. In that case, there will be some prismatic effect in all directions of gaze. When prescribing decentration of a lens to obtain prismatic effect, be careful to consider what happens in other directions of gaze.

What are Fresnel prisms? What adverse optical effects might a patient experience while wearing one?

They can be thought of as an array of very narrow adjacent prism stripes engraved onto a thin sheet of plastic. They are commercially available as Press-On prisms in powers of 1-30 PD. Press-On prisms are cut to fit the back of a spectacle lens, and are simply pressed on under water like a decal. Patients may note decreased visual acuity, especially with the higher power prisms, due to chromatic aberration and glare (Figure 16-11).

Figure 16-11. Comparison of conventional prism with Fresnel prism.

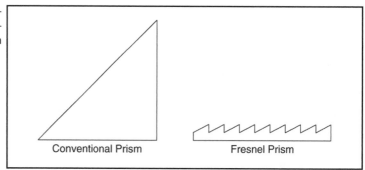

What aberration is the basis of the duochrome test? How does it work?

Chromatic aberration, which is caused by the prismatic effect of lenses, is the basis of the duochrome test. Different wavelengths are bent (refracted) to a different degree by the eye's optical system. Shorter rays (green, blue) are bent more strongly and red rays are refracted less so (Figure 16-12).

Figure 16-12. Chromatic aberration of a lens and of the eye.

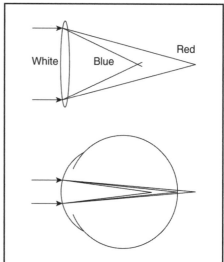

The chromatic interval between red and blue wavelengths is about 1.25 D. When perfectly focused for yellow light, we are myopic for blue light and hyperopic for red light. The duochrome (bichrome) test uses red and green filters to produce chromatic aberration of about 0.50 D between the focus of the left and right sides of the projected visual acuity chart.

How should the duochrome test be performed? Can it be used with a color blind individual?

The duochrome test is performed one eye at a time. To perform the test, begin from the fogged direction (over-plus the subject so that the letters are clearer on the red side). Now adjust the sphere until the letters are equally clear on both sides. At this point, the red rays are as far behind the retina as the green rays are in front. This means that yellow light will be in perfect focus—the optimal refractive correction for viewing in white light. Errors can result if accommodation is not relaxed or if visual acuity is poor.

Yes, the duochrome test can be used with "color blind" individuals. They have as much chromatic aberration as anyone else.

A spectacle-corrected 10-year-old soccer player with 8 D of myopia is referred by a plastic surgeon to "clear the globe" prior to repair of an orbital floor facture. The boy presented to an emergency room when he developed double vision after a head butt at a recent soccer game. The plastic surgeon plans to operate in 30 minutes for a white-eyed orbital blowout fracture. The patient appears comfortable but he is quite bothered by the double vision. Cover testing reveals a 16 prism diopter left hypertropia. The extraocular movements are full. Your review of the CT scan indicates a nondisplaced orbital fracture with no clear evidence of entrapment; the radiologist confirms this. The athlete's glasses are bent and crooked.

A. What happened and what should you do?

The bent glasses have displaced optical centers, which is inducing prism. Repeat the measurements with the correction in trial frames to confirm that there is no strabismus, and then cancel the surgery and repair the glasses!

B. Which way were the glasses bent to create the impression on cover testing that the right eye was hypotropic?

If the glasses were bent to displace the right lens up and the left lens down, then the right eye would view the world below the optical center of a minus lens, thus exposing the eye to base *down* prism. Meanwhile, the left eye would be viewing the world above the optical center of a minus lens (base *up* prism). These prism powers would sum up and may be considered to act as base down prism on the right. To neutralize this induced prism, you would use base up prism on the right, which you know from clinical experience is used to neutralize a right hypotropia (left hypertropia). Thus, the glasses were bent such that the right lens went up and the left lens went down.

C. How badly were the glasses bent (in cm) to induce a 16 PD left hypertropia?

If the optical centers were displaced by 1 cm over each eye, then the induced deviation is 1 cm x 8 D = 8 PD base down on the right and 1 cm x 8 D = 8 PD base up on the left, for a total of 16 PD of induced prism.

17

Instruments

What is a Geneva lens clock?

A Geneva lens clock is an instrument that measures the radius of curvature of a lens surface by the deflection of a movable pin. Most lens clocks are calibrated for crown glass (n = 1.523), so a conversion factor is necessary to use it with other lens materials. Take care not to scratch plastic lenses with the pins of a lens clock that is made for glass lenses.

Why is base curve clinically relevant?

Base curves affect the magnification of the lens, and most dioptric powers can be dispensed in a number of base curves. If a patient is accustomed to a certain base curve (magnification), a change in base curve—despite constant overall dioptric power—may cause asthenopic symptoms.

What principle is the manual lensmeter based on? How does it work?

The optometer principle. An illuminated mobile target is moved back and forth along the optical axis of an unknown lens (contact or spectacle) until the vergence leaving the lens is zero. The target is viewed with a telescope that provides magnification and relaxes accommodation, allowing precise detection of parallel rays at neutralization. Without the addition of a second lens, called a fixed (or "optometer") lens, a lensmeter would have a very nonlinear dial. That is, the dial would have to be turned a large amount when going from 1 to 2 D and a microscopic amount when going from 14 to 15 D. The fixed lens is carefully positioned so that its focal plane exactly coincides with the location of the unknown lens (Figure 17-1). The observer moves the illuminated mobile target back and forth behind the fixed lens; this varies the vergence entering the fixed lens. The linear distance of the target from the rear focal plane of the fixed lens is then directly proportional to the power of the unknown lens. If the power of the unknown lens is zero, the target will be in best focus when it is positioned at the rear focal plane of the fixed lens (see Figure 17-1).

Hunter DG, West CE.
Last Minute Optics: A Concise Review of Optics, Refraction, and Contact Lenses, Second Edition (pp 99-110).
© 2010 SLACK Incorporated

Figure 17-1. Optometer principle. "f" is the focal length of the fixed lens.

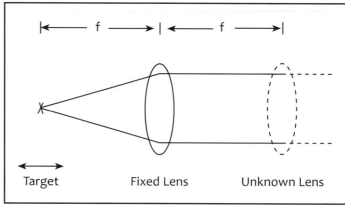

Target   Fixed Lens   Unknown Lens

How does one measure prism power and decentration of the optical centers in glasses?

Have the patient fixate on your opposing eye. Mark the glasses with a felt marker where the line of sight intersects the lenses. Place the glasses in the lensmeter with the mark centered in the nose cone of the lensmeter. If the lensmeter cross target is not centered, prismatic power is present. This power may be due *either* to ground-in prism or to decentration. If there is *no* position where the cross target is centered, then some of the prism is ground-in. If you can find a position where the cross target is centered, some of the prism is due to decentration, but there may also be some ground-in prismatic power, especially in high-power lenses. This is why it is important to mark the intersection of the line of sight with the glasses to measure actual prism power (as seen by the patient) accurately.

The amount of prismatic power may be read off of the scale in the lensmeter eyepiece. If the lensmeter cross target is deflected *down* to the circle marked "1," then there is 1 PD of base-*down* prism. If the cross target is off scale, you must use a hand-held prism with its base oriented in the direction opposite the prismatic power to determine the amount of power. For example, if the cross target is centered when a 10 PD prism is placed base-*up* in the lensmeter light path, the glasses must have 10 PD of base-*down* prism.

What are the origins of "with" and "against" movement during retinoscopy?

At neutralization, the far point of the eye coincides with the peephole of the retinoscope, and all of the light returning from the eye passes through the peephole to the examiner. Thus, the pupil seems to fill with light.

When the far point of the eye is in front of or behind the examiner, only part of the light returning from the eye passes through the peephole. Thus, the pupil is bright where the originating light rays pass through the peephole, and dark where the returning light is blocked, creating the appearance of a streak. If the patient's eye has too much plus power (far point between the observer and the patient), the light rays are inverted before they reach the examiner, the streak appears to move opposite to the illumination, and "against" movement is seen. If not enough plus power is present (far point beyond the peephole), the light rays are not inverted and "with" motion is observed.

What is optical doubling, why is it useful, and what instruments use it?

It is often desirable to be able to measure small distances in a living patient with great precision. Theoretically, it should be possible to simply engrave a reticle (scale) on an eyepiece and measure the object of interest (for example, the width of a magnified image of a vertically oriented corneal cross-section) (Figure 17-2, left). This is not practical, since a patient

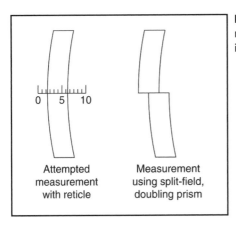

Figure 17-2. Optical doubling eliminates the need to align a stationary ruler with the moving image of interest.

must remain virtually motionless while the reticle is lined up with the magnified image. Instead of doing this, imagine that the field of view can be split into identical upper and lower halves. Now imagine that the lower half is shifted to the right with a prism (Figure 17-2, right). The prism power can be adjusted until the left side (epithelium in this example) of the lower image lines up with the right side (endothelium) of the upper image. With a doubling prism, the separation of the 2 images does not change even when the patient moves (since both images move together). The actual size (thickness in this example) of the image can be calculated from the power of the prism used to align the images.

Image doubling is used in keratometry (for both Bausch & Lomb keratometers, which use constant mire size and variable image doubling, and Javal-Schiotz keratometers, which use constant image doubling with adjustable mire separation), in pachymetry, and in applanation tonometry.

What does a keratometer *truly* measure?

A keratometer measures the size of an image that is reflected by the cornea. The size is measured precisely using optical doubling. The object (mire) is of a known size, and the ratio of object to image size is used to calculate the magnifying power of a small (3 mm) annulus of the cornea. The magnification is in turn used to calculate the radius of curvature (in mm) of the cornea using the formula $D = 2 / r$. But *reflective* power (power when light bounces off the cornea) is not the same as *refractive* power (power when light passes through the cornea). Since we know the radius of curvature we can determine the refractive power using the formula:

$$D = (n' - n) / r$$

where the keratometer measures r, uses 1.000 for the index of refraction of air, and 1.3375 as the index of refraction of the cornea. This translates the radius of curvature to diopters.

Note that we don't know that every cornea has a refractive index of 1.3375. This is known as the "standardized" refractive index for the cornea, a number that assigns exactly 45 D of power to a cornea with a radius of curvature of 7.5 mm. If you should ever have to do keratometry on a visitor from another planet, the refractive index of the cornea might be quite different, so you will only be able to measure the corneal curvature with certainty.

What other modalities are available for evaluating the corneal surface?

Placido's disk has concentric circles that are reflected by the corneal surface, and their regularity (or irregularity) is studied subjectively. The analysis of displacement of concentric circles is the basis of many computerized topographical mapping systems. These instruments evaluate the entire corneal surface, rather than a small central portion as with keratometry.

The videokeratoscope computer identifies the rings and measures the distance between each ring at multiple sites to construct a topographic image of the cornea.

The Pentacam corneal topography system uses a rotating, scanning optical slit to image the anterior segment, and then performs mathematical calculations to create a 3D model. This allows measurement of anterior and posterior corneal topography, corneal thickness, anterior chamber dimensions, and cornea and lens density.

What is wavefront analysis?

In wavefront analysis, a laser beam is projected onto the fovea, and the relected wave is analyzed. In a perfect optical system, the reflected wave should be perfectly flat. In reality, the wave is distorted by various aberrations in the optical system of the eye. These distortions are measured in reflected light by a series of sensors, which detect the angle and direction of these reflected waves. The output of the sensors is processed to develop a color map that represents the imperfections at every point in the reflected wave. Analysis of these maps can be used to plan the corneal ablation required for refractive surgery.

What basic principle was so revolutionary in the development of the slit lamp?

A *common pivot point.* A slit lamp has an illumination and a viewing system; the illuminated slit is imaged precisely over the common pivot point of the illumination and viewing paths. This allows the slit beam and the object of regard to remain in precise focus at all times.

How does a specular microscope work? What are normal "counts"?

By viewing directly into the reflected illuminating light, the interface between the aqueous humor and endothelial cells can be visualized and used to count the endothelial cells. A typical endothelial cell count for a young person is 3000 cells/mm^2, and for older adults, 2250 cells/mm^2.

What is a pachymeter for? How does it work?

Pachymeters measure thickness (corneal thickness or anterior chamber depth), and use either optical doubling (older models that attach to slit lamp) or ultrasound (ultrasound biomicroscopy) to accomplish it.

How does an applanation tonometer work? What do the numbers on the scale mean? Can errors be induced when measuring an astigmatic cornea?

See the previous discussion of optical doubling. The split-field plastic prism separates the half fields by 3.06 mm. Increasing the force against the eye expands the tear meniscus until the inner edges of the semicircles are aligned. The applanated area is 3.06 mm in diameter (not area!), and the force is read off the scale in *dynes.* Multiplying the force in dynes times 10 yields the intraocular pressure in mm Hg.

If significant astigmatism is present, the applanated area will be an ellipse rather than a circle. You can compensate by rotating the prism to align the red mark with the axis of the *minus cylinder* correction. Another accurate way to avoid error in astigmatic corneas is to applanate twice: once with the split prism horizontal, and once with it vertical. The true pressure is the average of the two readings. Astigmatic errors become measurable with more than 4 D of corneal cylinder, but even 4 D of corneal cylinder causes an error of only 1 mm Hg.

Goldie Mann is a 38-year-old woman who presents for a comprehensive eye examination 4 years after successful LASIK surgery. She has no complaints. Her 48-year-old brother was recently diagnosed as a glaucoma suspect, and her mother went blind from glaucoma. On examination, uncorrected visual acuity is 20/25 in the right eye and 20/20 in the left eye. Intraocular pressure measures 17 mm Hg in each eye using Goldmann applanation tonometry. The optic disks appear healthy, with a cup-to-disk ratio of 0.3. How do you interpret the intraocular pressure in this case?

The accuracy of Goldmann applanation tonometry depends on assumptions that corneal rigidity and corneal thickness are relatively constant among individuals. Patients who have had LASIK surgery have thinner corneas. As a result, the measured intraocular pressure may underestimate the true pressure. (Conversely, intraocular pressure may be overestimated in patients with thick or rigid corneas.) While there are nomograms available that relate corneal thickness (pachymetry) to intraocular pressure, they may not be particularly accurate in post-LASIK patients. In these cases, some experts advise to measure pressure off-center, away from the treated area. Others suggest using the pneumatonometer or newer intraocular pressure measuring devices. No matter how the pressure is measured, patients at risk for glaucoma need to be followed more closely if they have abnormally thin or thick corneas. In this case the pressures of 17 are probably underestimates, and the patient should have objective documentation of the optic nerve appearance and baseline visual field testing.

If you increase the magnification in any magnifying optical system, what happens to your depth perception?

In general, the axial magnification is the square of the transverse magnification, so any increase in transverse magnification is accompanied by a greater increase in axial magnification. The depth of the image will appear to increase more than the width. Therefore, the perceived depth of the object will be exaggerated relative to its width.

Describe how a binocular indirect ophthalmoscope works.

The binocular indirect ophthalmoscope has an illumination and a viewing system. Both the illumination and viewing paths pass through the hand-held condensing lens. A mirror places the light source closer to the examiner's eyes. The binocular eyepieces contain prisms that utilize total internal reflection to reduce the observer's interpupillary distance to about 15 mm. This allows both of the examiner's eyes, along with the illumination source, to be imaged within the patient's dilated pupil, giving the examiner a well-lighted and *binocular* view. An inverted aerial image of the patient's retina is formed between the condensing lens and the observer. The aerial image is viewed by the examiner (Figure 17-3).

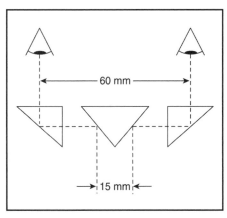

Figure 17-3. Eyepieces of a binocular indirect ophthalmoscope reduce the interpupillary distance from 60 mm to 15 mm.

How does a small-pupil indirect ophthalmoscope work? What is the tradeoff?

A small pupil scope decreases the interpupillary distance even further, and may move the light source closer to the eyes as well. This allows both eyes and the light source to be imaged through a smaller pupil. The tradeoff is decreased depth perception and increased glare from reflections. Note that if a small pupil indirect ophthalmoscope is not available, it may be possible to obtain a monocular view of the retina with an ordinary indirect ophthalmoscope *if* the light source on the head piece is close enough to your eyes.

What is the magnification of the image of an emmetropic subject's retina using a 20 D condensing lens?

The magnification of the aerial image of an emmetrope's retina is the ratio of the power of the patient's eye to the power of the condensing lens. Thus for a 20 D condensing lens, the transverse magnification is 60 / 20 = 3X. The axial magnification (how deep or high a fundus lesion looks) is the square of the lateral magnification, 3^2 = 9X. However, our interpupillary distance is reduced from 60 to 15 mm, reducing our stereoscopic clues fourfold, and this reduces the *perceived* axial magnification of a 20 D condensing lens to 9 / 4 = 2.25X.

Which gives the least distortion of depth during indirect ophthalmoscopy, a 15 D, 20 D, or 30 D indirect ophthalmoscope lens? Why?

A 15 D lens gives the least distortion. Use calculations similar to those in the previous problem (Table 17-1).

Table 17-1				
Calculation of Magnification and Depth Distortion				
LENS POWER	**TRANSVERSE MAGNIFICATION (60 / LP)**	**AXIAL MAGNIFICATION (TM²)**	**TRUE AXIAL MAGNIFICATION (AM / 4)**	**DEPTH DISTORTION (TRUE AM/ TM)**
30 D	2X	4X	1X	0.50
20 D	3X	9X	2.25X	0.75
15 D	4X	16X	4X	1.00
LP: Lens Power; TM: Transverse magnification; AM: Axial magnification; True AM: True axial magnification				

The 30 D lens flattens the image by 50%, while the 15 D lens provides equal transverse and depth magnification, causing no flattening of the image.

What is the magnification of an emmetropic person's retina when viewed with a direct ophthalmoscope?

The examiner uses the optics of the patient's eye as a simple magnifier: 60 D / 4 = 15X. That is, the patient's retina looks 15 times larger than it would if it were removed from the eye and held 25 cm from the observer's eye.

When viewed by an emmetropic physician, why does a myope's optic disk look bigger than an emmetrope's?

In myopic patients, the eye has more built-in plus power than it needs (it contains a plus error lens). The minus lens dialed into the ophthalmoscope in order to see the retina clearly forms the eyepiece of a Galilean telescope. The observer is thus looking at the patient's fundus through a Galilean telescope, providing the observer with a magnified image of the patient's optic disk. The opposite is true in a hyperopic eye.

What are the two main components of a surgical loupe?

A surgical loupe is a Galilean telescope combined with a near add. The working distance of the surgical loupe is determined by the focal length of the add.

What sorts of problems might one encounter if surgical loupes are not properly adjusted?

Problems that result from improperly adjusted binocular loupes include vertical and horizontal deviations secondary to induced prismatic effects, and accommodative problems secondary to improperly selected working distances.

Dr. Bellicose, a mid-level attending at a teaching hospital, arrives in the OR in an unusually anxious mood because he has purchased brand new, super-deluxe, 2X loupes, but he is worried that he will not like the change. Sure enough, before the end of the first case, Dr. Bellicose has a miserable headache. His resident, Dr. Gunner McSharp, dons the loupes, ponders a moment, then recognizes the problem. "Well Dr. Bellicose, it appears… yes… that the working distance of these new loupes is set at 40 cm, but you have been working all day at 25 cm!"

"Gunner," Dr. Bellicose howls, "even I know that if the loupes are off by that much, I might have to accommodate… what… a diopter and a half to see clearly at 25 cm. I may be older than you, but I can dial in 1.5 D of accommodation, no problem."

"Okay Dr. Bellicose," replies Gunner. "Considering that you are the Director of Residency Education, you must be correct. I apologize for my ignorance. You should send those terrible loupes back to the manufacturer."

Who is correct? Just how much accommodation has Dr. Bellicose been exerting to operate through his new loupes?

Just as a telescope multiplies image size through angular magnification, it also multiplies the vergence of light. For this reason, the accommodative demand when viewing a near object through a telescope is multiplied by the *square* of the angular magnification. In this case, the loupes are set at 40 cm (1 / (0.4 m) = 2.5 D), but the surgeon is working at 25 cm (1 / (0.25 m) = 4 D). Dr. Bellicose calculated that he should have to accommodate just (4 − 2.5) = 1.5 D to see clearly. But Gunner knows best, the true accommodation is found by multiplying 1.5 D times the square of the magnification (2 x 2 = 4), or (4 x 1.5) = 6 D.

Discuss the principles of fluorescein angiography.

Sodium fluorescein dye is injected into a vein, where 60% to 80% is bound to serum albumin, and 20% to 40% remains unbound. Fluorescein has a maximum absorption at 485 nm and peak fluorescence at 530 nm. A white flash from the camera passes through an interference filter, and blue light enters the eye. The fluorescein absorbs the blue light and emits a longer wavelength yellow-green (530 nm) light, along with the reflected blue light. A blue-blocking interference filter is placed in front of the camera to prevent the reflected blue wavelengths from entering the camera. The image formed by the yellow-green wavelengths is imaged on high-contrast black-and-white film.

How is indocyanine green angiography different from fluorescein angiography?

In fluorescein angiography, the largest portion of light energy during excitation and emission is absorbed by the RPE and macular xanthophyll, which can make it difficult to visualize certain choroidal patterns. Furthermore, unbound fluorescein leaks rapidly from the highly fenestrated choriocapillaris.

In contrast, indocyanine green (ICG) is highly (98%) bound to plasma proteins. ICG has a maximum absorption and peak fluorescence at 805 nm and 835 nm, respectively. Because this falls in the near-infrared part of the spectrum, the near-infrared light used more readily penetrates the RPE and macular xanthophyll, and the choroidal circulation is more easily visualized.

How does optical coherence tomography (OCT) work? Why is an OCT image so much more detailed than an ultrasound image?

In OCT, coherent, directional light is bounced off of the posterior pole of the eye, and as the light waves pass through the retina, they are slowed different amounts depending on the composition of the tissue. To detect the changes, a second coherent beam of light is split off of the first and bounced off of a mirror that is the same distance from the source as the retina (the "reference arm.") The phase and/or frequency of the two light varying beams creates interference, which is detected by combining the waves in an interferometer. The variations are computed to create an "A scan" type of image. As with ultrasound, a series of A scans are then combined to create a 2-dimensional B scan image of the retina. Because light has a much higher frequency (shorter wavelength) than sound, the OCT image has a much higher resolution than an A or B scan ultrasound.

What's so special about laser light?

Laser light is monochromatic (all photons have the same wavelength), coherent, polarized, and directional (non-spreading). (See Section 1, *Basic Principles*.) It is very intense light that is capable of delivering a lot of energy to a very small area.

What does "laser" stand for? What are the basic components of a laser?

Light Amplification by Stimulated Emission of Radiation. The 3 basic ingredients of a laser are as follows:

1. A power source to supply energy, think of this as *The Pump*

2. An active medium with special properties that can generate light energy by emitting photons, think of this as *The Filling*

3. A chamber with mirrors at opposite ends to reflect the energy back and forth, one of which is partially transmitting; call this *The Chamber*

To build your own laser, just put The Filling in The Chamber and pump it with The Pump. For example, put some helium and neon in a chamber. The chamber has 2 mirrors, one of which reflects all of the time and the other reflects 98% of the time but transmits 2% of the time. Now start "pumping" the chamber with, say, some rhythmic electrical discharges between electrodes. Each time there is a discharge, some photons are released from the gas. But they are not released from the chamber—they reflect back and forth, over and over between the mirrors an average of 50 times before they are finally released. Once they are released, the photons are all the same wavelength (thanks to your choice of The Filling), coherent (thanks to The Pump), and directional (thanks to The Chamber). *Voila*, you have a laser beam.

Fun fact: Exotic substances like argon and yttrium are typical photon emitters used as the active medium in lasers, but more common substances can also be induced to emit photons. In fact, edible lasers have been created from colored gelatin.[1]

How does laser light damage tissue? Name 3 distinct mechanisms.

1. Burn it! Thermal damage through absorption. Energy is absorbed by the tissue, there is local rise in temperature, proteins denature, and photocoagulation results. It follows that the tissue to be treated must absorb at the wavelength used to make photocoagulation possible.

2. Blow it up! Disruption/plasma formation (see Nd:YAG, described later).

3. Break it apart! Sublimation. The laser energy disrupts covalent bonds without heating or coagulating protein (see excimer laser, described later).

You are planning to use the thermal energy supplied by a laser to coagulate a choroidal neovascular lesion in an eye with cataract and previous vitreous hemorrhage. What wavelengths of laser energy are best absorbed by different layers of the eye?

- *Argon blue green (488 nm).* Absorbed by RPE, xanthophyll, iris, and hemoglobin

- *Argon green (515 nm).* Absorbed by RPE, xanthophyll, iris, and hemoglobin

- *Krypton yellow (568 nm).* Absorbed by RPE, choroid, and iris

- *Krypton red (647 nm).* Absorbed by melanin (iris, PRE, and choroid), and poorly by hemoglobin and xanthophyll. This gives krypton an advantage over argon when there is a vitreous hemorrhage or maturing lens

- *Diode (815 nm).* Absorbed solely by melanin in the pigment epithelial layer, with thermal damage to surrounding tissues. The retina can thus be treated either through the pupil or through the sclera, as minimal diode laser energy is absorbed by the sclera. As with Krypton, this is an advantage when there is blood in the eye. Theoretically, it may allow delivery of laser energy to a choroidal neovascular membrane without damaging overlying nerve fiber bundles

What does Nd:YAG stand for?

Neodymium:yttrium-aluminum-garnet. This type of laser creates 1064 nm infrared, nonvisible light. The aiming beam is a helium/neon laser with a wavelength of 633 nm. Because the refractive index for the eye is lower for longer wavelengths, the YAG will be posteriorly focused relative to the aiming beam. The YAG laser's high power pulses are used to cause optical breakdown and localized mechanical disruptions with the formation of plasma (ions and rapidly moving molecules). The YAG laser does not rely on absorption, so semitransparent membranes (eg, the posterior capsule) can be cut.

What range of light does the excimer laser emit?

Excimer lasers produce high-power ultraviolet radiation, ranging from 193 to 351 nm. At this wavelength, the light is absorbed by all layers of the eye, especially the cornea, which is the first thing (besides air and tears) that the light encounters.

When are CO_2 lasers used in ophthalmic practice?

CO_2 lasers utilize invisible 10,600 nm energy and vaporize tissue. They are usually used as a scalpel during ophthalmic plastic surgery procedures. The energy does not penetrate beyond a few microns.

Why do some lasers let you set power in Joules, and others in watts?

A Joule is a watt of energy delivered for 1 second. That is, if the time of energy delivery is known, the number of Joules can be calculated. With continuous wave lasers (such as argon), the time of energy delivery varies (depending on the shutter setting of the instrument), which makes it unnecessary and impractical to set power in Joules. Therefore the power setting is in watts. With pulsed lasers (such as Nd:YAG), the power is delivered as short bursts of energy, so the watts vary over time. The total energy in each pulse can be measured, and therefore Joules are a more appropriate unit of measure.

A continuous wave laser has a power setting of 1 watt. The spot size is decreased from 100 μm to 50 μm. How does the brightness (irradiance) of the laser beam change?

The tissue effect of a laser is related to its "brightness," as indicated by the energy density (Joules/cm^2) or irradiance (watts/cm^2). If the power is 1 watt and the area of the spot is 2 cm^2, the irradiance is (1/2) or 0.5 watts/cm^2. If the power is kept the same but the area of the spot is decreased to 1 cm^2, the irradiance is (1/1) or 1 watt/cm^2. Thus, for any given power setting, the energy density or irradiance increases as the spot size decreases. Intuitively, this should make sense—if you concentrate the same amount of energy into a smaller area (spot size), the brightness (intensity) should increase accordingly.

Note that we don't usually use lasers that have a spot size of 2 cm^2, but the same reasoning applies to spot sizes of 100 versus 50 μm. "Spot size" is usually specified in *diameter*, so a spot size of 100 microns covers an area of about 0.00008 cm^2, and a spot size of 50 μm. covers an area of about 0.00002 cm^2. Thus, a power of just 100 milliwatts delivers an irradiance of about 1.25 kilowatts/cm^2 or 1,250,000 millwatts/cm^2 if the spot size is 100 μm, or about 5 kilowatts/cm^2 if the spot size is 50 μm.

REFERENCE

1.   Hansch TW. Edible lasers and other delights. *Optics and Photonics News.* 2005;Feb:14-16.

18

Good People, Bad Optics: The Dissatisfied Patient

A 56-year-old man seeks consultation regarding diplopia. He had been well with no ocular or medical problems until 1 week earlier when he woke up with double vision. He contacted his brother-in-law, a neuroradiologist, for advice. The patient ended up getting a CAT scan, a PET scan, an MRI, and an MRA. Then his cat got a PET scan. All were negative. Someone suggested that it might be a good idea to see an eye doctor, and the patient came to you.

A. What key ocular historical points do you need to elicit in a patient with new-onset double vision?

Is there any history of patching in childhood? (This could reveal a recurrence of childhood strabismus or other abnormal binocular vision.) How sudden was the onset? (Gradual onset is less likely to be related to an acute neurologic insult.) Is the vision still double with one eye closed? (Monocular diplopia is often confused with binocular diplopia.) Does it change throughout the day? (Variable strabismus might suggest myasthenia.) How many images are there? (This can give another clue to monocular diplopia or visual issues not related to strabismus.) Are the images separated vertically, horizontally, or a combination of both? Is one image also tilted? (It can be difficult to separate oblique diplopia [combined horizontal and vertical] from torsional diplopia [tilted image.])

B. In this case, the patient says that the double vision goes away when he covers his left eye. You at first assume that this means he is having binocular diplopia, but cover testing reveals no detectable strabismus. During cover testing, he reports that the double vision disappears when the left eye is covered, but remains when the right eye is covered. You diagnose monocular diplopia in the left eye. He says that there are usually 2 images but sometimes 3. It is like a ghost image, always overlapping, sometimes more visible than others, especially on the television. What are the causes of monocular diplopia? Name a key diagnostic procedure.

Monocular diplopia is usually caused by irregularities in the refractive surfaces of the eyes. Causes from front to back include dry eye, corneal epithelial irregularities, stromal scarring, refractive surgery, uncorrected refractive error, and incipient cataract. Non-optical causes such as retinal irregularities are extremely rare. The key diagnostic step is to see whether the diplopia improves though a pinhole occluder, as optically-induced monocular diplopia is almost always reduced or eliminated by the pinhole. If it remains with the pinhole occluder

Hunter DG, West CE.
Last Minute Optics: A Concise Review of Optics, Refraction, and Contact Lenses, Second Edition (pp 111-114).
© 2010 SLACK Incorporated

in place, then the problem could be with the retina, or there may be more complex cortical processes including factitious visual loss. Another helpful test is retinoscopy, which can be used to subjectively assess for irregularities in the refractive surfaces of the eyes. Follow that with a refraction; correction of refractive error is often enough to reduce or eliminate the problem.

C.   In this case, the pinhole occluder removes the diplopia, but a careful refraction does not help. There is no evidence of dry eye—the tear film looks good, though there is mild blepharitis. What treatments can you offer?

Despite the lack of dry eye or significant blepharitis, you could try treating both in case the irregularity is subtle. In some cases, pilocarpine drops can help by shrinking the pupil and covering areas of aberration, but the side effects are generally prohibitive. If the problem is in the lens, cataract surgery will help, but might be more than is really necessary. In many cases, patients are satisfied with the explanation that the problem is optical and not a sign of a brain tumor, and no additional treatment is necessary.

Your next patient of the morning, Dee Plopia, is a 46-year-old woman who also presents with double vision. In this case, Dee thinks that the problem occurred either just before or just after you changed her glasses prescription 6 months ago, but then again maybe she has had double vision for years. You measure the new glasses and find:

OD: -2.00 -1.00 x 72

OS: -1.50 -0.50 x 93

Progressive add +1.25 D OU

Visual acuity with this new correction is 20/15 in each eye and J1+ at near. Her visual acuity wearing the old glasses was 20/30 in each eye at the last visit. The double vision is binocular. Cover testing with correction reveals a right hypertropia of 4 PD, worse in left gaze and worse with head tilt right. What additional test should you consider?

In this case it appears that the patient has a right superior oblique palsy. But why did she suddenly get worse? You need to check the old glasses. When you do, you discover that the old glasses actually had 3 PD of base-down prism on the right. You neglected to prescribe the same amount of prism when you gave the new prescription, because you never measured the old glasses for prism power. Other causes of strabismus secondary to bad glasses might include inadvertent addition of prism in the new glasses, too much minus or too little plus causing an accommodative esotropia, or a change in bifocal power causing increased accommodative effort at near.

Your next patient is Noah Win, a 65-year-old engineer who is even more dissatisfied with his new glasses (prescribed last month) than the last patient. Your first thought is that you must have changed the power or axis of his cylinder correction when he came in for his last refraction. You measure the old and new glasses, but to your surprise, the new glasses are *exactly the same power as the old glasses,* and give exactly the same visual acuity—better than 20/20 in each eye. What might cause a patient to be unhappy with new glasses that have the same prescription as the old, and how could you check for problems?

After checking for power, cylinder, and axis as noted in the question, check for prism. This may have been unintentionally added due to improper location of the optical centers. Compare base curves of the two lenses using a lens clock—some patients will complain when a new frame design requires a change in base curves, which can lead to alterations in the magnification caused by the new lenses. While you have the lens clock out, confirm that the astigmatic power was ground onto the inner surface of the lens in both cases. Look to see if the new glasses have a different vertex distance, or if they were damaged in some way causing them to settle differently on the patient's face. Evaluate the coatings and tints—maybe the anti-reflective coating was left off this time. Re-check the bifocal power and find out whether the style of add has changed— ie, flat top to round top, a change in segment height, or more subtly, a change in the brand of progressive lens.

You are now running 45 minutes behind, and your next patient is a 30-year-old woman, a new patient who cannot be refracted to better than 20/50 in either eye. Before commencing with pupil dilation and a medical evaluation for vision loss, what steps can you take to confirm that this is not a refractive problem?

First, recheck visual acuity through the pinhole *with your best refraction in place.* If vision does not improve, she probably has a medical problem. But if the visual acuity improves, there is still a refractive problem contributing to subnormal vision. Note that autorefractors may also be fooled in these cases; there is no substitute for checking pinhole acuity with best refraction in place. If vision does improve with the pinhole, try using a higher-powered, hand-held Jackson Cross cylinder to refine the subjective refraction. At 20/50, the patient does not have good enough vision to detect a change in sharpness when you flip the ±0.25 D Jackson Cross cylinder that is built into the phoropter. If that does not help, look for irregular astigmatism in the retinoscopic reflex, as noted above. If this is present, a contact lens over-refraction is in order. If visual acuity does not change with the pinhole, there are a few simple things to consider. Make sure that the eye chart in the room is properly calibrated and properly illuminated. Check visual acuity with both eyes open in case latent nystagmus is present.

You are now well over an hour behind schedule, and your next patient is a 38-year-old man complaining of headaches and vaguely describing trouble with his eyes. He has been wearing the same low myopic correction for at least 3 years. You suppress the urge to shout, "Headache? You want to talk about a headache? Well I have the worst splitting headache known to man and you aren't helping it one bit!" What refractive issues should you consider when evaluating headache and asthenopia in a patient?

The considerations for this patient are quite similar to those of the other patients you have seen this morning. Perform cover testing to look for phoria or tropia. Check the manifest refraction and near visual acuity. The key test in this case is going to be a cycloplegic refraction. The patient may have latent hyperopia, which will cause him more trouble as he approaches the presbyopic years; or he may be over-minused, which he used to tolerate fine, but now he is running into trouble as his accommodative amplitudes decline over time.

19

Important Formulas

THE MOST BASIC FORMULAS

FOCAL LENGTH

$$f = \frac{1}{D}$$

f = focal length in meters
D = lens power in diopters

VERGENCE FORMULA

U + D = V

U = vergence of rays entering lens (object rays)
D = vergence added by lens (lens power)
V = vergence of rays leaving lens (image rays)

PRENTICE'S RULE

PD = h x D

PD = prism diopters
h = distance from optical center in cm
D = diopters of lens power

SPHERICAL EQUIVALENT

$$SE = Sph + \frac{1}{2} Cyl$$

SE = spherical equivalent
Sph = sphere
Cyl = cylinder

Hunter DG, West CE.
Last Minute Optics: A Concise Review of Optics, Refraction,
and Contact Lenses, Second Edition (pp 115-118).
© 2010 SLACK Incorporated

MAGNIFICATION

Retinal Image Height

$$\frac{\text{object height}}{\text{retinal image height}} = \frac{\text{distance from nodal point}}{17 \text{ mm}}$$

Spectacle Lens Magnification

$M_{\text{spectacle lens}} = 2\%$ per diopter of power
(Assume 12 mm vertex distance)

Transverse Magnification

$$M_{\text{transverse}} = \frac{\text{image distance}}{\text{object distance}}$$

Indirect Ophthalmoscope Lens

$$M_{\text{indirect ophthalmoscope}} = \frac{D_{\text{eye}}}{D_{\text{lens}}} = \frac{60}{D_{\text{lens}}}$$

Axial Magnification

$M_{\text{axial}} = M^2_{\text{transverse}}$

Simple Magnifier

$$M_{\text{simple magnifier}} = \frac{D}{4}$$

D = lens power
Standard reference distance = 1/4 m (0.25 m)
For nonstandard reference distance, replace 1/4 m with new reference distance

Telescope

$$M_{\text{telescope}} = \frac{D_{\text{eyepiece}}}{D_{\text{objective}}}$$

POWER FORMULAS

REFRACTING POWER OF A SPHERICAL SURFACE

$$D_s = \frac{(n' - n)}{r}$$

D_s = refracting power of surface
$(n' - n)$ = difference in refractive index
r = radius of curvature of surface
To determine sign: use imaginary rectangle

POWER OF A THIN LENS IMMERSED IN FLUID

$$\frac{D_{air}}{D_{fluid}} = \frac{(n_{IOL} - n_{air})}{(n_{IOL} - n_{fluid})}$$

D_{air} = power of lens in air
D_{fluid} = power of lens in fluid
n_{IOL} = refractive index of lens
n_{fluid} = refractive index of fluid
n_{air} = 1.000

REFLECTING POWER OF A SPHERICAL MIRROR

$$D_{reflecting} = \frac{1}{f} = \frac{2}{r}$$

$D_{reflecting}$ = surface reflecting power
f = focal length
r = radius of curvature

IOL POWER (SRK FORMULA)

$$D_{IOL} = A - 2.5\,(L) - 0.9\,(K)$$

D_{IOL} = Recommended power for emmetropia
A = A constant
L = axial length (mm)
K = keratometry reading (Diopters)
For desired ametropia, change IOL power by 1.50 D for each diopter of desired ametropia.
For short or long eyes, see Table 11-1 for A constant modification (SRK II).

Index